T0185763

Space Modeling with SolidWorks and NX

Jože Duhovnik · Ivan Demšar
Primož Drešar

Space Modeling with SolidWorks and NX

 Springer

Jože Duhovnik
Ivan Demšar
Primož Drešar
Engineering Design and Transport System
University of Ljubljana
Ljubljana
Slovenia

ISBN 978-3-319-38143-5   ISBN 978-3-319-03862-9   (eBook)
DOI 10.1007/978-3-319-03862-9
Springer Cham Heidelberg New York Dordrecht London

Printed on acid-free paper

Springer is part of Springer Science+Business Media (www.springer.com)

Preface

This book is for the reader who wants to acquire new engineering skills related to concrete, complex solutions through the modeling of objects in 3D space. Anyone who has ever dealt with modeling, or just simple drawing, soon finds out that the straightforward examples from exercises are easily repeatable. However, there are few such examples in real-life. Therefore, we have decided to offer a better and more detailed presentation of the complex shapes and problems associated with the modeling of real objects. Based on our years of experience with a large number of people through graduate and master's study programs we have decided to opt for the gradual acquisition of knowledge, which leaves it well grounded in the student.

The extent of the material exceeds that required for regular study. However, there are a couple of universities that deal in detail with such basic knowledge, the basic language of engineers, i.e., the ability to present new products and ideas. Our cooperation with industry, various institutes, and our long experience in designing products with various complexities have resulted in a wide range of examples compared to regular monographs. We leave the readers to form their own opinions about the large number of presented examples.

We should point out that the transferring of dimensions and details to technical documentation represents the pinnacle of modeling. Of course, anybody dealing with the quality and perfection of products is aware of this. Should the readers find that the presented knowledge about detailed modeling and the special forms of technical language are both new and interesting, and that they are able to generalize and use it confidently, we will derive great satisfaction.

During the Bologna reforms, our colleagues suggested that in the first semester students should acquire new knowledge about descriptive geometry and technical drawing. In the second semester, this knowledge is upgraded with skills related to space modeling. At the end of the semester, the students are made familiar with transferring a shape to the high-quality representation of a product by means of technical documentation. Our 8 years of experience has shown that in later stages—in machine elements, energy systems, and manufacturing engineering courses—students can easily present their ideas using complex models. This was a proof of our intention to also include knowledge about the high-quality presentation of ideas in the 3D environment in a new profile of engineering competences.

Developing methods and the complexity of models for industrial design, for example, requires familiarity with free-form surfaces, which will be the next logical step after the content of this book. We believe that the presented systematics will allow both students and their teachers to easily recognize the upgrade to their existing knowledge. This belief was also confirmed by our staff engineers, who were willing to attend special courses to upgrade these skills, mainly in 2D space.

The hand-sketching chapter was taken from the book "Engineering Graphics," where a co-author is Prof. Milan Kljajin from Strossmayer University (Osijek, Croatia). He confirmed that we were able to use this content. We would like to express many thanks to him for this permission. Our book would be incomplete without this chapter.

A book of this size cannot be written by a single author. A substantial contribution to its final form came from a team of Ph.D. students, Damijan Zorko and Pavel Tomšč. Mateja Maffi, Simon Demšar, and Paul McGuiness took care of the translation and proofreading, and the technical editing was carried out by Janez Krek. They all deserve sincere gratitude for their tolerance and patience.

Jože Duhovnik
Ivan Demšar
Primož Drešar

Contents

Chapter 1
Introduction

Abstract Various layouts are presented for the working places in a classroom that allow communications between the tutor and the students. This is followed by a presentation of the different forms of engineering graphics that are subject to changes in the Research and Development (R&D) process. This R&D process includes three loops in the research and strategic parts, and a golden loop in the development phase. The developments of the graphic expression from the sketch to the working documentation are specified, and directly connected with the phases of the R&D process.

A knowledge and understanding of space and its dimensions are important for both engineers and designers. Some people are born with a talent to understand space, objects and their distribution, while others can learn it. However, such a talent must be developed, which makes learning about and understanding the concept of space vital, even for the most talented of individuals. Learning provides a solid basis for a detailed study of space, particularly the systematic organization of space. Excellence in terms of knowledge is acquired through constant training and the resulting improvement. Should engineers or designers not learn about the fundamental forms of expression, such as a sketch on a piece of paper and/or a digitised model on a screen, they will gradually lose the ability to present their ideas with clarity.

  Without training, they forget that it would be better to replace their gesticulating and gobbledygook with a pencil and paper in their hands. Of course, instead of a pencil, it is often enough to take a piece of clay brick, and some concrete or a rock for the writing surface. With a sketch it only takes a few lines to present our thoughts clearly. At first sight, this method of presentation might seem too rough to create new ideas; however, it is a fact that an initial idea is often presented with very rough concepts. An idea that originates in the abstract, conceptual world of an engineer represents creativity that is later reflected in the real-world environment. This is the virtue of having an engineering understanding of the world, as each line, each thought, expressed by a sketch, provides the reliability of connections. Together with numerical proofs it provides the possibility for modern man to be able to live in tall buildings and to travel by aeroplanes, trains and ships. It is difficult to imagine a person using a flat in a skyscraper or flying in aeroplanes with a feeling that the skyscraper might fall apart or the aeroplane might not take off because, in such a technical system, an object would not have been specified in all the required details.

The objective of a space-modelling course is to acquire the practical engineering and designer skills to express 3D shapes and the necessary variations by assorted means of expression. The exercises during this course provide training that will reinforce an individual's knowledge. The course is suitable for the first year, second semester of a university study programme, in order to provide a student of technology or design with the fundamental knowledge of expressing and specifying a shape in space, similar to the need for a solid basis in mathematics, physics, materials science.

It is possible to gain an understanding of space modelling immediately after gaining the basics of descriptive geometry and technical documentation. The lectures and exercises take place weekly. They are structured in such a way that the students upgrade their knowledge from simple objects to complex models. The whole course of exercises includes three interactive knowledge assessments. Such a course of exercises requires quality software and computer equipment. Special attention has been paid to this area as it is not possible to acquire high-quality knowledge without high-quality hardware and software. Good computing equipment provides a good response during the use of individual routines, while high-quality software provides reliability for the performance of the modelling.

Particular knowledge is presented through examples of characteristic shapes and features. Each characteristic shape is presented with a short introduction, followed by presenting the use of the feature in the manufacturing process. General description is presented, which is useful for all modelling software. In the next two sub-chapters we describe modelling with SolidWorks software, which is more user-friendly for beginners. This is followed by NX, which belongs to the area of advanced modelling software. Our objective is not to teach the routines for a particular software package, i.e., SolidWorks or NX in our case, but to teach the modelling philosophy with the use of features from different shapes that are more or less the same for all modelling software. The differences between the different modelling-software packages lie mainly in the user interfaces.

In preparing the work programme for basic modelling we have included all the necessary skills, knowledge and findings that complement the need for acquiring a comprehensive knowledge of solid-body modelling. Table 1.1 shows the topics that a student at a good university will cover (2 h per week) and the knowledge that he or she can acquire with additional study and training at home after 30 h of exercises.

From the presented programme it is clear that there is a special emphasis on the individuality of the exercises. This means that it is not appropriate for two students to do the exercises behind a single screen or on one computer. It goes without saying that the structure of the classroom and the software for systematic exercises require a special approach, i.e., a special architecture for the room. High-quality software, integrated into a unit, is also vital.

Table 1.1 The topics and the order of tutorials for space modelling

Excercise no.	Topic covered during exercises	Note
1.	Role and meaning of modelling for engineering	Introductory tutorial, Boolean algebra, sketching
2.	Making a sketch and extrude	
3.	Making a sketch and revolve	
4.	Auxiliary shapes of modelling	Fillet, chamfer
5.	Assessment—combined models	
6.	Complex shapes—sweep, loft	Sweep, loft
7.	Welded structures	
8.	Sheet-metal products	
9.	Measuring a physical model, digitising, parameterisation	
10.	Assessment—physical models	
11.	Assemblies of structures with models—bottom-up technique	
12.	Assemblies of structures with models—top-down technique	
13.	Documentation—assembly drawing	
14.	Documentation—working drawing	
15.	Assessment—examples as finished units	Seminar work

1.1 Arranging Rooms for the Study of Modelling

Organizing a suitable room for a small group is an important task. Exercises can only be performed well if at least one assistant (a leader and an assistant, who could be a student from a higher grade) is available for every 6–8 students. A suitable group has between 12 and 16 students. With two experts in the room for an accelerated interaction with the students, the access to the students should be fast. This is achieved by providing at least 60–70 cm of space behind the seats for the expert to gain access to each student.

Larger classrooms with up to 30 places are possible; however, the work then slows down significantly, despite having up to twice as many experts. In this case there are four experts in each classroom (and three in special cases). The classroom plan for 16 places is shown in Fig. 1.1. A classroom with 32 places is shown in Fig. 1.2. In the case that the classroom is narrower and longer, the plan can be designed in accordance with the template in Fig. 1.3.

Because a programme-integrated environment is vital for a quick response and good communications with the students, it is shown on its own in Fig. 1.4. The concept involves a joint server with 16 working places and the main specifications of the hardware that were available on the market between 2010 and 2014.

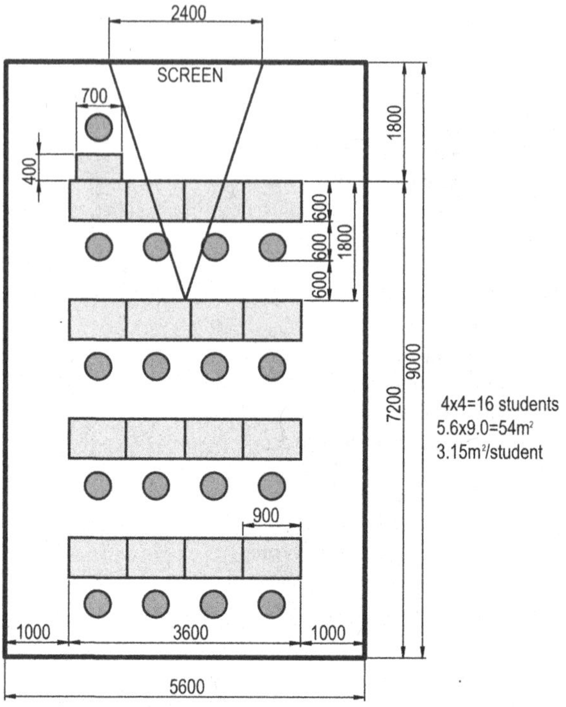

Fig. 1.1 Classroom plan for 16 working places and connections

1.2 Development and Design Process and Engineering Graphics

Engineering uses different demonstration techniques: writing (describing), mathematical formulae (equations) and drawing (sketching, modelling). Writing is a matter of general human education, whereas mathematical formulae are a way to use equations in order to describe natural phenomena. Engineering drawing is used to show nature in space. Artistic drawing, on the other hand, serves a different purpose. The latter describes things as realistically and as clearly as possible, whereas engineering drawing in its final form yields a plan. This plan should be as clear to as many users of different cultures and skills as possible, and was the basis for the introduction of the standards for engineering expression, i.e., technical drawing. Such engineering expression through drawing is generally referred to as engineering graphics.

The more sophisticated is a product, the more detailed is the knowledge of engineering graphics that needs to be applied. While thinking, in the abstract world, when a rough, immaterialized idea is emerging, we should begin with a freehand sketch. Freehand sketching uses specific techniques and requires a certain level of knowledge. The techniques of freehand drawing are based on manual dexterity, and on refined motoric functions of the hands and fingers. For this reason it is vital to learn

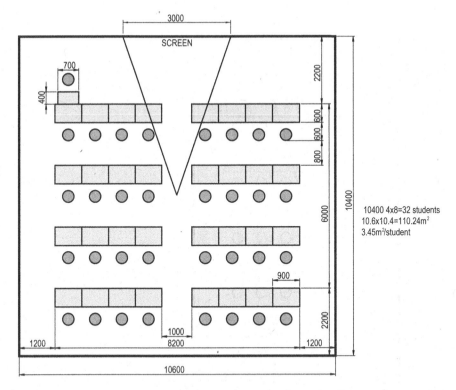

Fig. 1.2 Classroom plan for 32 working places and connections

freehand-drawing techniques so that our records will be clear and unambiguous, in a similar way to clear writing.

Unskilled engineers, not using a sophisticated freehand-drawing technique, will express themselves with a lower quality. It is difficult for them to articulate an idea; it is difficult to place it in space. With freehand-drawing techniques it is vital to take account of the abilities in proportional ratios, in both details and dimensions.

As engineers, we develop and study how to make people's lives in their surroundings easier in practical terms. And so, developing the world, we think how to work smoothly in a process that allows us to materialize an idea with a product in the form of an object. In this view we have developed different approaches and procedures, one of them being the R&D process. In this process, individual phases are specified. They are then processed consecutively, depending on the expected results of the task during each phase. As a rule, they are not limited in terms of detailed analyses and researches in a particular phase, nor by going back to the previous phase for further information. The R&D process is known as iterative, supplementing and generally based on the decisions of the team or its decision makers. What to expect as a result in each phase is specified in descriptive, analytical-mathematical, and graphic forms. Generally, there are several forms of R&D processes as well as different methods

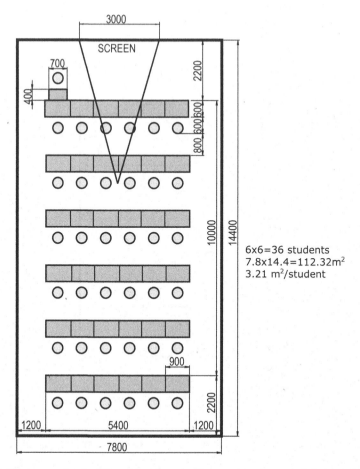

Fig. 1.3 Classroom plan for 32 working places and connections (an extended classroom with a screen of the same size)

of processing. In this book we will elaborate on one that was developed in both theory and during work on projects. The model is not limited to a traditional industrial product. It is generalized and includes all the necessary synergy of knowledge. The R&D procedure was first published by Duhovnik [7] as the Golden Loop in the Design & Development process in 2003 at TU Delft and University of Wroclaw. It was then supplemented with early-phase research on the technology of processes [12] that a product should provide (Fig. 1.5).

At this point the focus will be on the main parts of the R&D process. More information and a more detailed explanation are available in the literature.

The innovative loop is of key importance for recognizing a problem and placing it in space. In the innovation loop we specify the task, the requirements and the wishes that play a part in defining a problem or a technological process (e.g., cutting

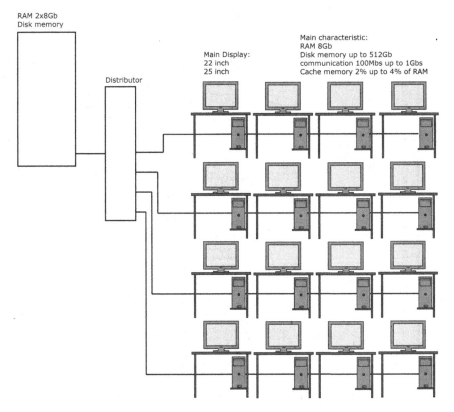

RAM 2x8Gb
Disk memory

Distributor

Main Display:
22 inch
25 inch

Main characteristic:
RAM 8Gb
Disk memory up to 512Gb
communication 100Mbs up to 1Gbs
Cache memory 2% up to 4% of RAM

Fig. 1.4 Network lecture room with hardware capabilities

bread, lifting a liquid and taking it into the mouth—a spoon is intended for the same process in both Chinese and European cuisines). All rough starting points are generally supported by rough sketches, rough models (Plasticine, real models made out of wood or virtual—digitised). Together with specifying the shape of an expected object, its function—described by parameters—should also be defined. However, not all parameters are of the same importance. The important parameters are identified by means of various methods. They are called the winning parameters. The winning parameter of each function is the one that significantly affects the fulfilment of a particular function, so it is called the winning parameter of the main function (it can be physical, chemical, biological, etc.).

The system-engineering loop places an expected result (innovation, a high-quality product)—vital for the materialization of a product—logically into space and time. An innovation that is not system-defined in the engineering, general, social environments, generally receives no response from the market or users, which keeps such an invention at the level of a hobby.

The application research loop is important in order to define the parameters that provide an optimum choice of function in the material, the conceptual and the user's

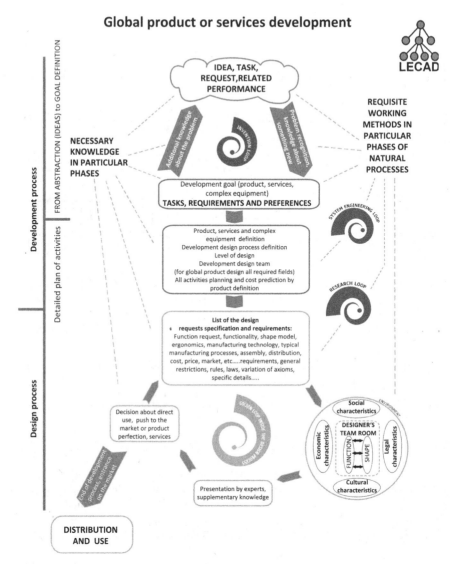

Fig. 1.5 The R&D process includes four loops [12]

environment. Fundamental and application researches are in tune with and of vital importance for the development of mankind. Supported by engineering solutions, they are the key to the reliable, high-quality and long-term development of mankind. Any deviation from the integrity and the connection between fundamental and application researches is a clear sign of the level of the development of a particular environment.

The golden loop in the development process defines a product in all its details, supplements fundamental researches for all key parameters that define the details and quality materialization of a product. These are referred to as application researches, and they are vital for defining the perfection of a product. Without detailed application researches—dealing with the interaction between the basic process and the material environment—a product cannot be sustainably and well placed in space and time. It was for this reason that it was termed the "golden loop".

With such a recognized R&D process it makes sense to determine the role of engineering graphics. It is up to the engineer to chose the forms of engineering graphics to be used during the individual phases of the R&D process.

At the beginning of development, man expressed quantity in the form of analogue notes and tried to reproduce them. However, there has always been a desire to digitise analogue records. Therefore, besides text digitising (in the form of ASCII code) at an accelerating rate due to its simplicity, there were increasing attempts to digitise the notes. At first, dot recognition was used; however, high-quality records take up a lot of memory. Together with dot recording, there were also attempts to describe sketches with vectors. Vector graphics found their way into engineering sketching, and recording specific shapes, including free, random areas, takes up significantly less memory.

Vector expressions are presented in various books and form the basis for the development of mathematical relations in all modelling software, including SolidWorks and NX. For reasons of simplicity and accessibility for a larger number of users, this book expresses engineering shapes using commands that will become standard in the communications between man (the user) and computer. The mathematical basics will be briefly presented in the third chapter. In engineering graphics it is our objective to quickly acquire a digitised record of a developing product, which is then translated into other types of analyses and simulations. So, we will skip transferring shapes from one software package to another, e.g., from modelling software into FEMs (finite-element methods), CFD (computational fluid dynamics) or Mold-flow (Simulation Moldflow software provides simulation tools for injection-mould design, plastic-part design, and the injection-moulding design process).

A sketch should be digitised as soon as possible and presented in 2D or 3D space. When we have rough, less-accurate data in the sketch, the product model (Fig. 1.6) is rough, too.

We can then proceed to a rough product model, or a wooden model, clay, hard polystyrene or Plasticine. In this case, 2D or 3D scanners are used to capture their dimensions, which are then transferred into modelling software in a digital form (Fig. 1.7).

Prior to the system engineering loop we need and expect to develop significant engineering details with rough calculations of the loads and deformations. Text and mathematical equations are generally used to describe the functions for such analyses. For shapes, it is a freehand sketch or sketching in a digital model. With modelling software, there are different approaches to sketching detailed shapes, which are then logically introduced into the basic digitised model. A shape, presented with a model, forms the basis to establish a link between the function and the shape in the research

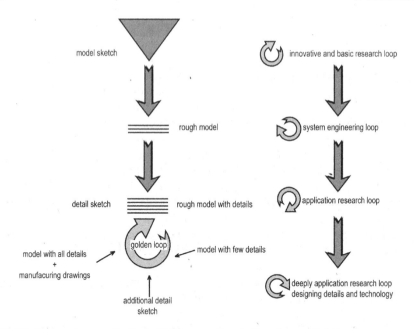

Fig. 1.6 Engineering graphics in the R&D process for the four characteristic loops

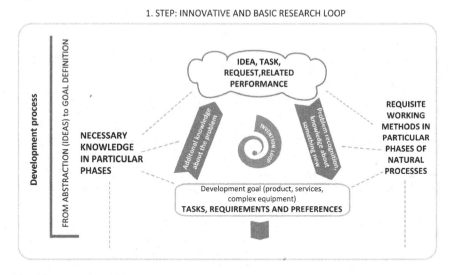

Fig. 1.7 Research and innovative loop and its detailed activities

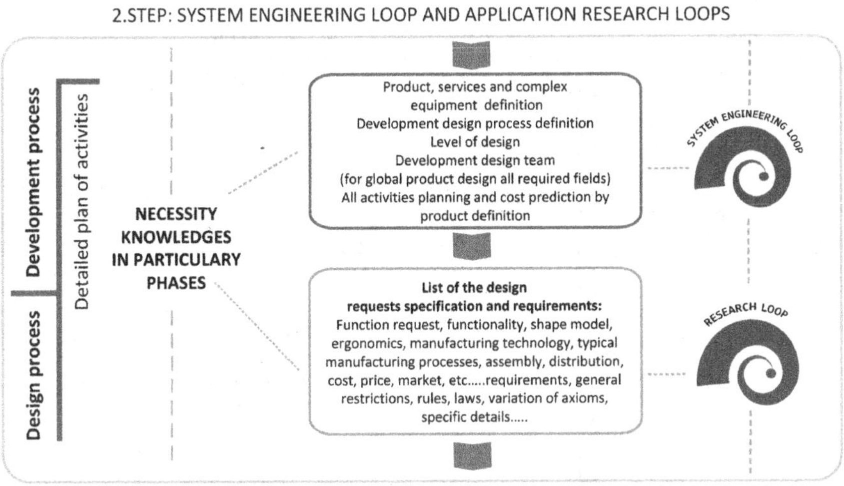

2.STEP: SYSTEM ENGINEERING LOOP AND APPLICATION RESEARCH LOOPS

Fig. 1.8 The system-engineering loop and the application research loop

loop. As an aid to clear presentation, the graphics use diagrams, where the relations between the obtained results, trends, approximations and predictions are established.

In both phases, i.e., the system-engineering phase and the application-research phase, a more accurate graphic model is presented. Important details are specified. During these two phases, freehand sketching is used only as an addition, not as a basis for the general modification of the model. The sketching of details is used for a better understanding of the shape. Data transfer in the basic form of the product is no longer analogue, as it is too rough. The details are improved by means of creative procedures, when an idea from the abstract world is presented to the environment. Handy sketching is used to improve the details of a model. Sketches in the form of an analogue record only serve as an aid for understanding the specific shapes or the product development. Figure 1.8 shows both phases and the type of engineering graphics used to improve the image of a product.

The use of engineering graphics can be presented with the example of a simple element of a shaft during the whole R&D process, shown in Fig. 1.9. On the left, there are four typical R&D process phases, while on the right, there are typical graphic representations that support the description and the mathematical formulations that are necessary for an analysis and a comprehensive presentation.

The whole engineering knowledge is presented with working documentation that represents the key information about the product in all its details. This presentation can only be made in the golden loop, where all the details are presented (calculations of the function and the shape) (Fig. 1.10).

When processing the information, it can be established that during the R&D process the product the details complement one another, which makes it easier to define a digital product model. At the beginning, the model is not burdened with the

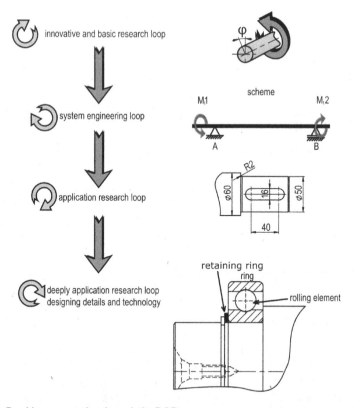

Fig. 1.9 Graphics presentation through the R&D process

details that are yet to be specified (e.g., radius, chamfer, etc.), but once their values have been calculated and verified by tests, they are inputted into the product model.

The final product shape is formed in the golden loop, where the design process is supplemented with detailed application researches. During this phase, the product is equipped with typical manufacturing and surface-treatment technologies. At this level, the digital model is accurate up to the values of dimensions specified by the technological process, which exceed it by 10^{-2}. During this phase the information about the shape is not entered by means of a sketch, but only by means of the dimensional value for a specific shape. If the details are even more complex, they should be presented in 2D graphics, in a conventional engineering method with a working drawing. The pinnacle of defining—which includes the specifics and the perfection of the production and manufacturing environment—is the working drawing. This makes the working documentation the most valuable industrial property relating to a product. Together with other documents it requires the highest level of confidentiality.

For the purpose of this book it has been decided to use values with an accuracy of 10^{-7} m. This means that we have deliberately decided to use numerical values

3.STEP: GOLDEN LOOP (IN-DEPTH APPLICATION RESEARCH) AND DESIGN ACTIVITIES

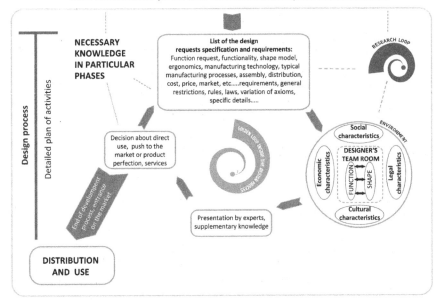

Fig. 1.10 A closed golden loop R&D process uses in-depth application researches in the function and technology of manufacturing, as well as the ergonomics-related shape

with two more decimal places than the positioning accuracy of working machines or measurements. The accuracy of working machines is at the level of 10^{-6} m or 0.001 mm. We are pointing this out in order to understand where and why the accuracy of geometry should be applied and improved, as the technology of measurement devices is ten times more accurate than that of working machines.

Chapter 2
Technical Freehand Sketching

Abstract Freehand sketching is important for the initial transfer of an idea from the engineer's abstract world. Sketching techniques, in 2D and 3D, are presented, together with details of the motoric functions of the hand during the sketching. A special section covers the phases of making a sketch for a working drawing.

2.1 Sketching Basics

The language of graphics in the engineering environment consists of a number of images and symbols. It is used for the daily communication of ideas and concepts. Each idea is verified by an attempt to express it with different concepts and in details. This is followed by precise technical drawings of products, which are required for their manufacture.

The basic ideas and concepts are usually outlined by means of freehand sketching. Using the criteria of technical drawing, symbols and characters, such a sketch becomes clear to anyone familiar with the international technical language. To avoid problems with the presentation of a technical system or product, it is vital to take account of all the criteria that are necessary for a technical drawing, i.e., a plan. The only difference lies in the fact that on a plan, everything is precisely drawn (including writing), whereas the accuracy of a sketch depends on the accuracy of its author and his or her talent. In life, talent can be developed, but not without work. Knowledge, however, can be acquired through learning and studying. A routine is established through work, similar to getting fit in sport. It makes sense that professional designers and engineers are skilled in the art of sketching. Generally, however, their sketching abilities are near perfect. Sketching allows them to communicate abstractions from the metaphysical part of their brains as a document for themselves and others.

Sketching is also used so as to be able to quickly present, in a graphical form, an idea that is related to a particular problem. Sketching significantly improves the communications between the members of a team, the drafter and the customer. A sketch is usually freehand, without any assistance or the use of instruments or aids that

J. Duhovnik et al., *Space Modeling with SolidWorks and NX*,
DOI: 10.1007/978-3-319-03862-9_2, © Springer International Publishing Switzerland 2015

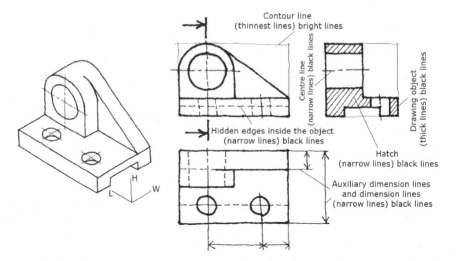

Fig. 2.1 A freehand drawn plane and a spatial sketch of an object, including the description of commonly used thicknesses, i.e., line intensities [10]

are otherwise used for technical drawing. Sketches can also be an integral part of analytical computations, where they are used to present and define the characteristic parameters of analysed physical objects.

Sketches are used to present the outside appearance of an object, with a little emphasis on concealed surfaces and features, which are included in the sketch in order to make the presentation as clear as possible. Sketches are direct graphical communications, often drawn in changeable conditions, such as at a building site, in a workshop or at a business meeting (Fig. 2.1).

In the field, a sketch can nowadays often provide a very good connection with photographs, showing nature or its details. By doing this, it is possible to include a sketch directly into the photograph of the environment in order to present an idea more clearly. By sketching, you can often present a technical system, force and power transmissions and other technical properties that are specific and focused on presenting certain information. In doing this, the information is 'cleared' of any redundant data that prevent a clearer insight into the subject of the analysis. For this very reason, engineers with inferior sketching abilities will face serious problems with communications in their development and research environment. As a result, it will take them longer to present an idea and it will be difficult for others to understand the details.

In this chapter we will explain and give guidelines for effective sketching. We will present the technique, process and templates showing how to make realistic sketches for engineering applications. Later, the technique of sketching will be presented together with short explanations that are related to the theory of technical drawing.

It should be pointed out that the greatest advantage of sketching is that it can be used anywhere and at anytime. You only need writing instruments (a pencil), drawing paper and a rubber (often not necessary), to make a sketch. With a finger, you can sketch in nature: on snow, sand or clay. It is possible to sketch with a piece of charcoal or a clay brick on concrete or asphalt. You can use any soft mineral in a colour that is different from the sketching surface, but it should be harder in this case.

With computer modelling, sketching is used in order to be able to present a rough idea or a modelling requirement. A sketch in this case is not precise, but specific details are exposed, as it is the image of the model that serves as a basis for an accurate digitised technical drawing model on the screen or plots, made on a plotter. Today, sketching is used—together with computer modelling as a way of providing high-quality space digitising—as an important part of the process of presenting the shape of a product or space in general.

2.2 CAD and Technical Freehand Drawing

Computer modelling (using CAD software) is possible if all the dimensions and features of a structure are defined at the input. Incomplete data do not allow a presentation, the object is recorded neither in RAM (on the screen) nor in the archives (file in data storage). This means that before modelling on the computer it is vital to acquire all the data as a free record, a sketch. Due to the volume of data, their structure is generally presented on a handy piece of paper. Generating a random model is conceptualized with a specific generic model, which the modelling software uses in its own way. For simple or less complex models or features, different types of software usually have similar shape generators. For the reasons of continuous data input, or at least to prevent lengthy breaks, you can take advantage of a rough presentation on a sketch that was done previously on paper. Later on, when the model is becoming increasingly complex, you should make a temporary copy of the model or detail on paper, which is then complemented with sketches that have improved details, which are necessary for the high-quality processing of a problem. Such a process is shown in Fig. 2.2.

Understanding data processing in freehand drawing can define the trends in the future development of computer modelling. Generally used modelling software (CAD programmes) still has difficulties with special shapes in terms of a computer description of specific technical products. For such products and for specific shapes, such as car tyres, steel structures (bars, framework structures, etc.), unfolded sheet metal, molecular structures etc., special modelling software is often developed. To present details, cross-sections, connecting systems, etc., specific routines are applied for each of those examples. In terms of future modelling software development we can expect the increasing use of natural communications, typical of human beings, such as: sketching, hands and fingers in space; eyesight, eye-pupil movement and image sharpening; hearing, communication for the simulation of sound phenomena; speech, voice commands for the computer; smell, recognizing the results of simulated processes.

In all these cases a representation in solid space is, of course, taken into account.

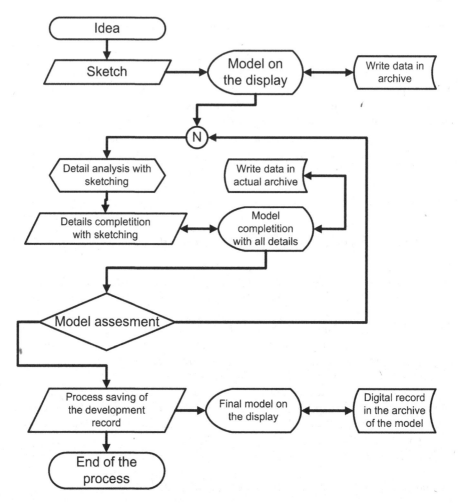

Fig. 2.2 Modelling with modelling software and computer equipment

2.3 Basic Rules of Freehand Drawing

2.3.1 Material, Sketching Tools

The sketching material consists of drawing and sketching tools. Drawing tools are required to make a sketch, and sketching tools offer the possibility to make a drawing.

Besides the basic material tools it is important to bear in mind the drawing technique. First, you need to learn how to make lines of different thickness.

Drawing tools Drawing tools consist of the various materials where sketches can be made. Technical sketches are usually made on opaque paper in sizes A4 or A3 (A2,

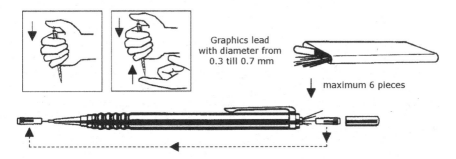

Fig. 2.3   A thinline propelling pencil for leads from 0.3 to 0.7 mm

A1 or A0 are also possible). Opaque paper is usually white, with weights of 0.80, 0.85, 1.00, 1.20, 1.40 and 1.60 N/dm^2 or more. Sketches on such paper are originals and can be copied on regular copiers. Continuous stationary paper is an extended version of the basic sizes. Besides opaque paper, tracing paper (the so-called paus paper) or drawing foils can be used in some cases.

Sketching tools Sketches are usually drawn with pencils in a black colour. However, in special cases, pencils with different colours can be used. Graphite pencils with different hardnesses, i.e., 9H—hardest, HB—medium hardness, 9B—softest). A graphite pencil of 2H hardness is normally used as it allows the drawing of both light (narrow) and dark (wide) lines by simply pressing the pencil against the drawing paper.

The term "propelling pencil" is often used. It refers to a pencil with a replaceable lead. For light thin lines, 2H grade leads are used. They are used to sketch construction lines, dimension lines, different signs, borders, and to write letters of up to 5 mm in size. 2B or B grade lead is used to draw contours, visible edges, shades and writing letters of 7 mm and more. Modern graphite pencils, used for technical graphics, are referred to as thinline propelling pencils. A thinline pencil of 0.3 mm generally used for the construction lines. Thicknesses between 0.5 and 0.7 mm are normally used to draw contours, i.e., dark heavy lines. Thinline pencils do not require any sharpening as the lines are of the same width as the width of the graphite lead (Fig. 2.3).

So, the following tools are recommended for sketching:

- two propelling pencils (with soft 2B or B and medium-hard 2H leads). Exceptionally, a third propelling pencil of HB grade (lead diameter 0.5 mm) for large sizes or drawings. For small drawings, a 4H lead is used with a diameter of 0.3 mm,
- a rubber,
- two set-squares (30°–60°–90° and 45°–45°–90°);
- a compass,
- white opaque paper (A4 or A3 size).

Line thickness It is possible to controllably draw all lines of thicknesses between 0.1 and 1 mm with a single thin lead of 0.7-mm thickness.

Straight lines should be drawn in one go, without breaks.

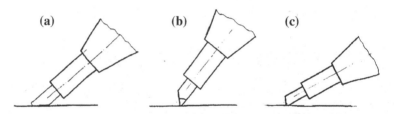

Fig. 2.4 Achieving lines of different thickness [4]. Different line thickness: thickness around 1 mm (**a**), thickness around 0.1 mm (**b**), thickness between 0.3 till 0.6 mm (**c**)

A thickness of 1 mm can be obtained by leaning the pencil and flat-guiding the lead. The lead should be leaned and guided along the surface without rotating it. Flatness can be obtained by a gentle slide against sandpaper (Fig. 2.4a).

A thickness of 0.1 mm can be obtained by guiding the lead on its tip, gently rotating the tip or the whole pencil. Due to wear, the line, i.e., the trace behind the lead, will of course be increasing in size. This can be solved by sliding the pencil against another paper in order to get the same sharp edges on the lead again. Pencils of 2H grade or more will maintain a thin line for longer (Fig. 2.4b).

The tip of the pencil can also be preserved for short fine (thin) lines. Without pressing too hard, the trace will be a lighter grey colour.

For the best copying and faxing results it is important for the lines to be black, which is achieved by pressing the tip hard enough against the paper. The drawing surface should be firm enough to prevent tears and impressions from the lead (Fig. 2.4c).

2.3.2 Sketching Straight Lines

For a clearer introduction to sketching systematics, the whole procedure will be shown for the example of making an arc and a straight line. Such exercises should be continuously revised to prevent a break in motoric function control and any loss of memory with respect to how to establish suitable conditions for drawing ratios.

Wrong: When the forearm rotates it functions as a compass (Fig. 2.5a).

Correct: Only the upper arm should rotate (Fig. 2.5b). The forearm and the hand stay still. The drawing hand should always be drawn in the direction of the body. Remember: pulling is mechanically always more stable than pushing. The easiest way to check the straightness of a line is to draw a line, pointing from the middle of our head towards our nose. Thus, the line is positioned at the centre of our line of sight.

Holding the pencil while sketching. The sketching trace shows the accuracy and skills in accurate rendering by the author of the sketch. For this reason, it is important to know how to achieve drawing accuracy more easily. The little finger and the

(a)   **(b)**

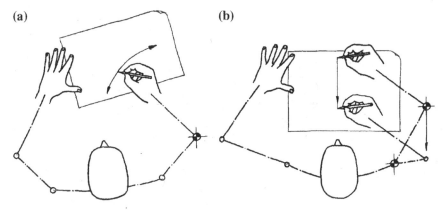

Fig. 2.5 Incorrect and correct movements of the arm when drawing *long straight lines* [4]. Forearm (**a**), upper arm (**b**)

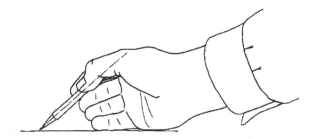

Fig. 2.6 A dry and clean fist should rest on the paper with a large area [4]

continuing edge of the hand should rest on the paper. All the other fingers, except the thumb, which partly holds the pencil, rest on the little finger (Fig. 2.6).

The pencil should point away from the finger tips (thumb and index finger) by around 40–60 mm in order to reach the paper surface in its entirety. In this case, the pencil should be held with the thumb, middle finger and index finger. The end of the pencil should be supported in such a way that the thumb and index finger are slightly bent. In this case, the pencil should have a straight end.

A long distance between the pencil tip and the fingers creates a gentle and steady pressure on the lead, which should be provided whenever fine or grey lines are being drawn. This position of the hand provides an obscured view of the area around the drawn lines (Fig. 2.7). It is important to prevent or minimize—with a sufficient spacing between the lead and the hand—any overlapping of the existing lines.

Some people do not use this grip due to the specific anatomy of the whole arm, especially the hand. However, there are cases when an improper introduction to the technique of sketching results in a cramped clutching of the pen or pencil. One should be aware that the hand muscles cannot remain tense or flexed for a long time. In both cases, it is a matter of partial breaks or a reduction in blood circulation. The first consequence is trembling in the back of the body, followed by a sudden

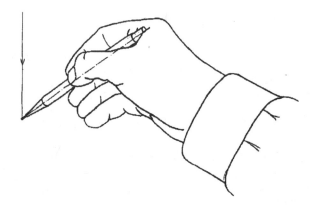

Fig. 2.7 This grip provides an obscured view of the area around the drawn lines [4]

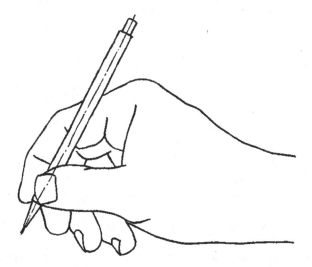

Fig. 2.8 This grip allows a very dark line. The black colour comes from a high pressure being applied to the surface [4]

muscle release in the entire hand musculature. This causes sudden uncontrollable hand movements, which leads to the drawing of undesired lines on the paper.

Line drawing Sketching should begin with fine grey lines, defining a coordinate system, a centreline, the edge of a drawing, etc. This is followed by defining the shape of the product (work, object) that is to be presented. Fine grey lines are still used here. When defining an object it can be presented with rough, thick lines, and the procedure begins with drawing heavy black lines. With each part of the sketch, the object comes to the forefront from the drawing plane. The thick lines are very black; however, they are heavier and of different types. The pencil should be held at its tip, straight and firmly pressed to the paper (Fig. 2.8). The distance between the finger tips and the pencil tip decreases, and is between 20 and 35 mm.

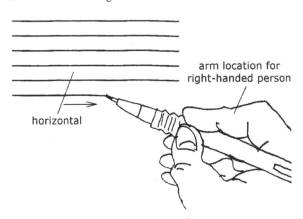

Fig. 2.9 Freehand sketching of *horizontal parallel lines* [10]

The hand should rest comfortably but firmly on the drawing plane. In this case, only the fingers are used to move the pencil, as moving the hand over the existing lines (grey, thin) could smudge it. For this reason, it is necessary to occasionally stop the process of drawing heavy, thick lines. Once the basis has been well presented and designed with thin grey lines, the presentation of the final shape can only become more accurate, compared to what was presented with the grey thin lines.

All this applies to people who use their right hand for sketching. Left-handed sketchers should mirror the kinematics and motoric functions of the hand over the centreline of the body. There is a mirroring problem due to a lack of testing and verifying the understanding of basic sketching elements. Creating freehand sketching rules should follow the same process as learning to write with the left hand.

Sketching straight lines Straight and curved lines are normally used for sketching. Straight lines can be divided into horizontal, vertical and oblique (sloping) lines. A standard approach to the freehand drawing of horizontal lines is from left to right (Fig. 2.9) for the right-handed, and from right to left for the left-handed. The freehand drawing of evenly spaced horizontal lines requires a fair deal of skill.

Attempting to draw straight lines often results in arcs or curved lines. This happens when the forearm remains in a fixed position, which is referred to as a stiff arm. So, when drawing horizontal lines, it is vital to move your whole arm. This is particularly important when sketching long horizontal lines. Your hand rests with its edge on the paper (Fig. 2.11) and the pencil sticks approximately 40 mm from the hand. Each straight line is drawn with the thumb and the index finger towards the body. Your movement must be controlled, and by changing the position of your arm, you can also draw oblique (sloping) lines.

You can also sketch long straight lines by drawing a number of short straight lines and then connecting their ends (Fig. 2.10). The third and very efficient method is the technique of freehand drawing of straight lines. This is executed by using the little finger as a support, sliding along a guide, represented by the edge of the drawing board. You should put the pencil on the starting point and then draw the line while

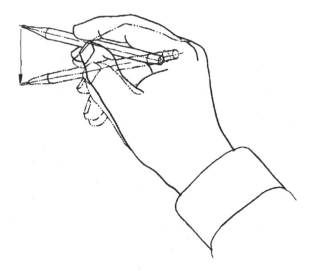

Fig. 2.10 Freehand sketching of *vertical parallel lines*, using the little finger as a guide to maintain the vertical direction (this is also possible for horizontal lines) [10]

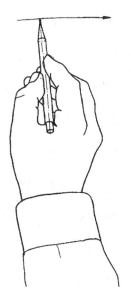

Fig. 2.11 Holding the pencil when drawing *horizontal straight lines* [4]

focusing your eyes on the end point. Figure 2.12 shows the procedure for the freehand sketching of vertical parallel lines, using the little finger as a guide to maintaining the vertical direction. When drawing horizontal lines, the little finger slides along the horizontal edge of the drawing board. You can also simply rotate the paper by 90°.

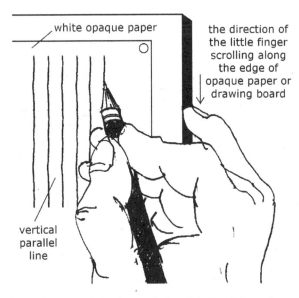

Fig. 2.12 Holding the pencil when drawing *vertical straight lines* [4]

Sketching vertical lines Vertical lines are usually drawn from the top to the bottom of the paper. The freehand drawing of evenly spaced vertical lines (Fig. 2.13) requires a lot of experience. However, the edge of the drawing board and the little finger can be of some assistance (Fig. 2.12).

Sketching horizontal lines When sketching horizontal lines, the hand and the forearm should pivot at the elbow. The thumb and the index finger can compensate for the forearm pivoting, while the edge of the hand should slide along the paper or travel above it (Fig. 2.12). You can first try drawing a line 'in the air', followed by an actual attempt, applying light pressure on the lead against the paper.

When drawing a medium-sized rectangle, the paper can be positioned in an orthogonal position (Fig. 2.15). With the paper rotated, vertical straight lines and horizontal straight lines are then drawn as oblique lines. The order of drawing the vertical and horizontal lines of a medium-sized rectangle is shown in Fig. 2.17.

Sketching oblique lines The right-handed person usually draw sloping or oblique lines towards the edges or from the bottom left to the top right of the page. It is rather unusual, even slighty difficult, to sketch an oblique line stretching from the top left towards the bottom right edge. The sketching is made much easier by rotating the drawing paper into a position that suits you better. Each motoric function of the hand corresponds to its own most suitable position of the paper (Figs. 2.14 and 2.16).

You can also draw straight lines by sketching them in the direction of the nose. To do this, rotate the paper according to the required angle for the oblique lines. In fact, you will be drawing a vertical line, relative to the drafter (Fig. 2.11). Beginners are advised to do a number of exercises for drawing long parallel lines on A4 or A3

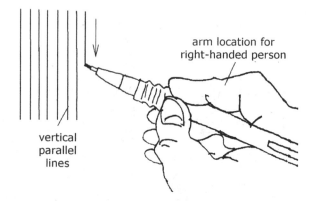

Fig. 2.13 Freehand sketching of *vertical parallel lines* [10]

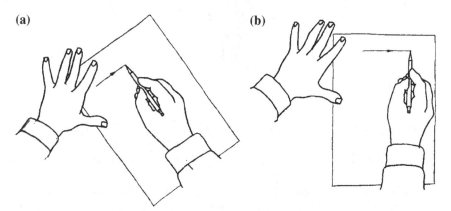

Fig. 2.14 Freehand sketching of *oblique parallel lines* [10]. Using forearm: long line (**a**), short line (**b**)

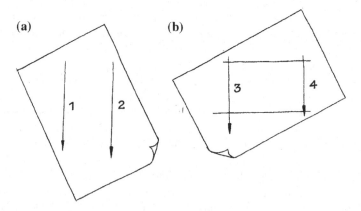

Fig. 2.15 Paper position for drawing [4]. Drawing larger rectangles or square: parallel line no. 1 and 2 (**a**), perpendicular line no. 3 and 4 (**b**)

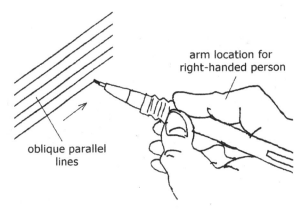

arm location for
right-handed person

oblique parallel
lines

Fig. 2.16 Freehand sketching of oblique parallel lines as vertical by drawing vertical lines along paper rotated in the direction of the nose (**a**), and exercises on A4 and A3 paper sizes (**b**) [4]

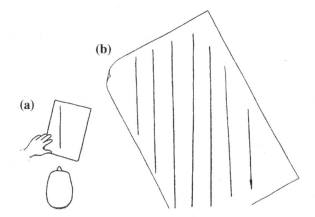

(b)

(a)

Fig. 2.17 Order of drawing *rectangle* edges—without hand position [4]. Long line should be controlled with two eyes in the middle (**a**), many lines we sketching compared straightness between them (**b**)

paper sizes (Figs. 2.13 and 2.14). In all cases, the paper is rotated so that the actually drawn vertical line is in fact oblique.

Drawing rectangles A rectangle (or a square) is a common shape in technical drawing. For sketching purposes, different techniques and rectangle sizes are used. They can be: large (exceeding 50 mm), medium (20–50 mm) or small (up to 20 mm).

With large rectangles, a technique similar to drawing long straight lines is used, combined with paper rotation (Fig. 2.15). You should become well acquainted with the procedure on A4 or A3 paper sizes. Drawing large rectangles is demanding (body control, paper rotation).

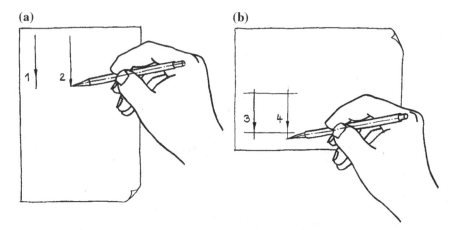

Fig. 2.18 Order of drawing a medium-sized rectangle—with hand position [4]. Drawing medium size rectangles or square: parallel line no. 1 and 2 (**a**), perpendicular line no. 3 and 4 (**b**)

Medium-sized rectangles (between 20 and 50 mm) are a very common feature of technical drawing (Fig. 2.18). They are not as difficult to sketch (body control, paper rotation) as large rectangles. The rotation of the paper can be limited. In the case of changing the sketching procedures or the drawing direction, the drawing technique should also be changed occasionally.

Small rectangles (under 20 mm) are drawn without rotating the paper and the hand. Only the thumb and index finger move (Fig. 2.18). During the drawing, the hand should rest still on the paper. The described thumb and index finger motion allows the drawing of lines of up to 20 mm in length (exceptionally up to 35 mm).

Everything said so far about drawing rectangles also applies to squares, as their angles are the same and the procedure is identical (Fig. 2.19).

The technique of drawing small squares also applies to the shapes that emerge when, for example, oblique lines tangentially meet the radii of fillets (Fig. 2.20). Fillets are first dotted (the distance between the dots is up to 2 mm), and when the straight lines are drawn, the fillets can be thickened.

Geometric shapes without right angles, such as a triangle, a pentagon, a hexagon or any other multi-angle shape, are freehand sketched, following the drawing technique that is used when sketching tools (a set-square, a ruler and a compass) are available. In principle, a shape is designed in the same way as when using the tools according to the rules of descriptive geometry, following the principle that all thin supporting lines are generally grey.

Fig. 2.19 Drawing small rectangles and other shapes with a fixed fist (only the thumb and index finger move) [4]. Drawing small rectangles or square: thumb and defined the line direction into the hand (**a**), direction into the hand (**b**)

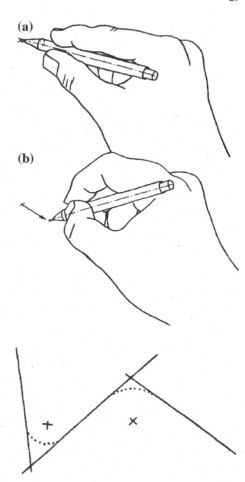

(a)

(b)

Fig. 2.20 A *dotted shape* before the final shape is drawn [4]

2.3.3 Sketching Curved Lines

Sketching curved lines Curved lines are either circles, arcs or irregular curves. To sketch a circle more precisely, draw the centreline first and mark it with radii. Shape the radii marks into a box, inside which you can sketch a circle.[1] Sketch the top left part of the circle first (draw the pencil in an anti-clockwise direction), followed by sketching the bottom-right part of the circle (move the pencil in a clockwise direction (Fig. 2.21b).

Larger circles require a second pair of centrelines, rotated by 45° relative to the first ones (Fig. 2.21c), due to an increase in the required number of radii marks from four to eight. For example, if you need four arcs (Fig. 2.21a), rotate the drawing paper for the last two quadrants, and sketching an arc is similar to sketching a circle.

---

[1] This procedure is known as the "boxing" of an illustration or a shape [4].

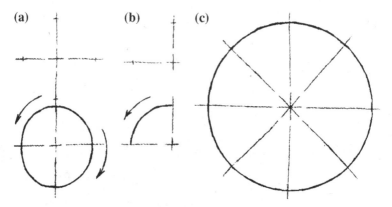

Fig. 2.21 Sketching a circle and an arc [19]. Principles: two arcs for small circle (**a**), sketching arc into clockwise direction (**b**), larger circle with eight arcs (45 degrees) (**c**)

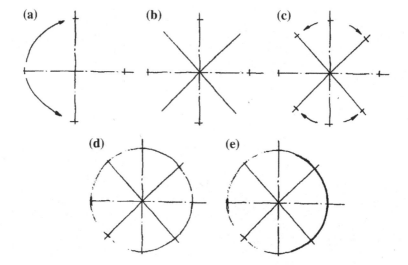

Fig. 2.22 The centre method's five steps for the freehand sketching of a circle. Steps: two perpendicular centre lines (**a**), rotation for 45 degrees perpendicular centre lines (**b**), radii dimensions transfer to all eight centre lines (**c**), first arc sketching (**d**), finalization with all arc the circle line (**e**)

The orientation and size of the arcs depend on the skills of the drafter. For medium and large arcs, it makes sense to draw centrelines and radii marks.

Sketching a circle is also referred to as the "centre method" (Fig. 2.22). It is similar to the above-described procedure in Fig. 2.21, with the difference being that the parts of the circle are sketched gradually, i.e., part by part. If the circle is of a short radius (under 50 mm), you can begin with the first phase and procedure from Fig. 2.23, while for larger circles (exceeding 50 mm) you should begin with the second phase.

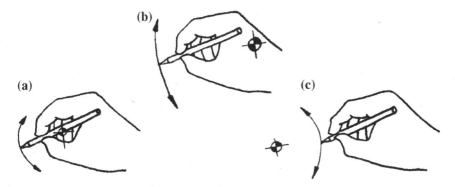

Fig. 2.23 Right (**a**, **b**) and wrong (**c**) positions of the fist, relative to the centre of the fillet when drawing a circle and an arc [4]

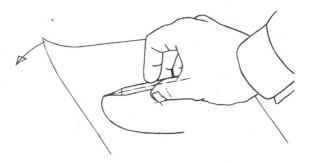

Fig. 2.24 Hand as a compass [4]

The hand can be used as a compass to draw lines of radii between 50 and 200 mm and their corresponding arcs (Fig. 2.24). Lean the part of the hand closest to the wrist against the paper, hold the pencil with the thumb and index finger and hold it in a position that corresponds to the required radius.

For all the movements on the paper, the hand should be clean and dry. Rotate the paper anticlockwise with your left hand, while the pencil and the hand remain motionless. This method allows the drawing of concentric circles. As a compass for marking dashed concentric circles or just circles, you can use a folded piece of paper with the appropriate marks, as shown in Fig. 2.25 or 2.27. Such thin circles can then be freehand thickened.

Sketching irregular curves Irregular curves are those without a specified radius of curvature. Sketch them with the strokes of a pencil that suit you best. Begin by drawing very narrow or light lines to sketch the background to produce the exact shape, followed by drawing wide and heavy lines for the final drawing of an irregular curve.

Fig. 2.25 "Paper compass" [4]

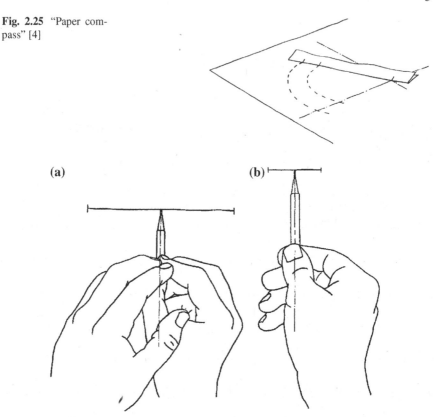

Fig. 2.26 Symmetry, judged with the fist and a pencil [4]. Long line with both hand (**a**), short line with one hand (**b**)

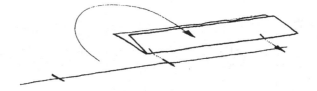

Fig. 2.27 Carrying an identical distance with paper [4]

Sketching grids Some drawing papers include printed sketching grids of narrow, usually light-blue, lines that can be of great assistance to drafters (Fig. 2.27). They usually come in the shape of a square or an isometric grid. A special case is a grid of thin, light-blue lines or dots, printed on transparent paper or a special drawing foil, which is very useful for drafters. Using the grid, it is easier to make clearer, freehand sketches.

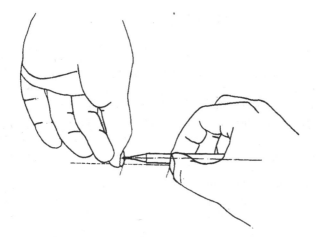

Fig. 2.28 Carrying an identical distance with a pencil and the *left* thumb [4]

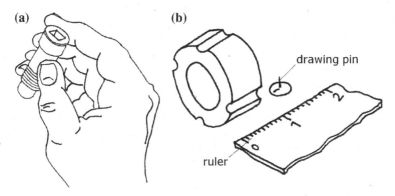

Fig. 2.29 Relative proportions of small objects can be shown by drawing them in the hand (a), or next to a familiar object, or a ruler (b) [10]

When you need to carry an identical distance and to maintain proportionality more easily, you can make use of a piece of paper with a mark on it (Fig. 2.27). Another option is shown in Fig. 2.28.

Sketching small objects with the hand and recognized proportions Small objects are sketched by taking them into your hands, and—by comparing their dimensions—carrying out proportional dimensions. Sketches often fail to provide an impression of the size of an object, especially when sketched without their dimensions. An impression of the size can be given by showing a hand, holding the object that is being observed (Fig. 2.29a). Instead of a hand, you can use other recognizable objects, such as a pencil or a ruler (Fig. 2.29b).

2.4 CAD and Technical Freehand Sketching

Having acquired the requisite sense of space, the designer can create the image of an object in his or her head, in abstraction. Based on a clear image in the mind, the idea is transferred onto the paper. When lacking the necessary feeling, a beginner should first sketch different models. The reader of the sketch should also acquire a sense of projection and recognize the rules of technical drawing and the sketching of realistic models. Those who use a sketch for modelling should acquire the wholeness of the model from the sketch. They should recognize the object in a 3D environment.

The sketching of models is, therefore, a skill shared by all engineers. Sketching is a process where individual phases follow one another in a particular order. But by not following individual sketching phases, gaps remain in the presentation space, i.e., there are indeterminacy of proportionality, inappropriate orientations of some shapes and the poor orientation of the lines of the product itself. Each phase of the process is therefore of great help in managing the sketching space. Below we will present a procedure for freehand sketching that is used for 2D product presentations. It is applied in the order shown in Figs. 2.30, 2.31, 2.32, 2.33, 2.34, 2.35, 2.36, 2.37, 2.38 and 2.39 [35].

Sketching procedures can be different. They are recognized by the different numbers of tasks or phases. For an easier understanding, they will be presented in one table, where the characteristics of individual phases are described (Table 2.1). The best known is procedural sketching with ten working phases (Figs. 2.30, 2.31, 2.32, 2.33, 2.34, 2.35, 2.36, 2.37, 2.38 and 2.39). The phases are precisely defined, which makes the procedure better suited to a beginner. A shortened procedure, known as stroke sketching, suggested by some other authors, has only five phases (the figures,

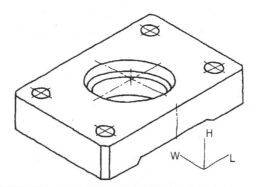

Name and main dimensions	Pos	View definition	Ortho. proj. by view and present.	Space present.	Simetric	Material
Basic plate 75 x 50 x 15	1		f.e. T	I	2x	S355J2G3

Fig. 2.30 First sketching phase. Figure shows a product and a handy working plan table

Fig. 2.31 Second sketching
phase [10]

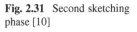

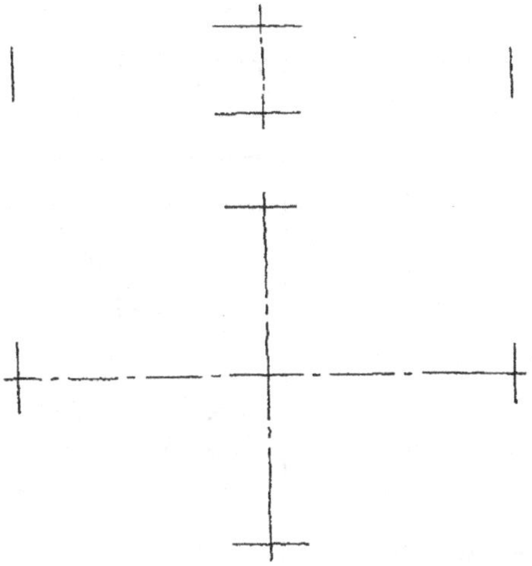

Fig. 2.32 Third sketching
phase [10]

linked to the procedures from Table 2.1). For both procedures it is important never
to erase the supporting or construction lines. Erasing or wiping is used in sketching
only in the case of drawing mistakes. In no case should erasing be used for the lines
that are procedurally required to determine the sketch.

Fig. 2.33 Fourth sketching
phase [10]

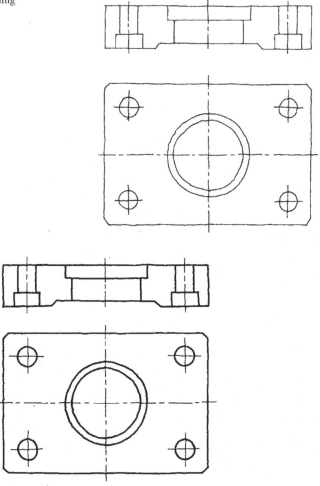

Fig. 2.34   Fifth sketching phase [10]

2.4.1 Procedural Sketching

Procedural sketching is used when dealing with an object that requires a detailed
specification of all the details because of its size and complexity.

First sketching phase (Fig. 2.30) is intended to specify the positions of the objects
to be drawn and the number of required projections to verify the possible symmetry
of the object and to decide whether to draw a cross-section and how the dividing
planes will run. All these data should be input into the working plan of the technical
documentation (planning table, Fig. 2.30).

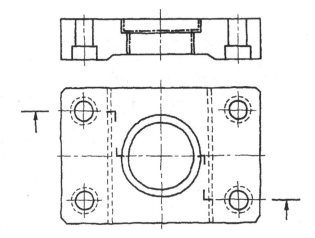

Fig. 2.35 Sixth sketching phase [10]

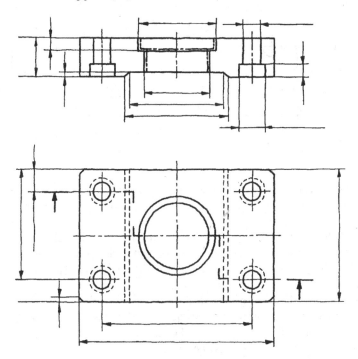

Fig. 2.36 Seventh sketching phase [10]

Second sketching phase (Fig. 2.31) is to verify the size of the space where the objects will be drawn, in order to be able to put all the projections, planned for the first phase on the same piece of paper or the selected format. It requires choosing a unit or deciding the size for 1 cm and specifying the dimensions of the object for

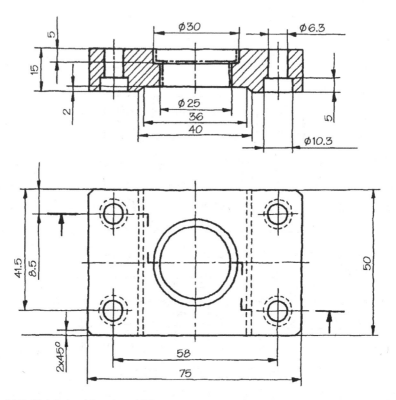

Fig. 2.37 Eighth sketching phase [10]

each projection, whereby you should take account of the dimensions that are visible in particular projections. You should save space for the dimensions and the drawing projections with a sufficient distance between them.

Third sketching phase (Fig. 2.32) is drawing the main bisectors with a hard pencil in all the projections where the object is symmetrical.

Fourth sketching phase (Fig. 2.33) begins with drawing the shapes of the object with thin lines. Start drawing the object with bisectors pointing outwards, dividing it in your mind into the basic geometrical shapes. Begin by drawing all the visible contours and edges in all the intended projections, drawing in each projection the most important part of the object, followed by the details.

Fifth sketching phase (Fig. 2.34) begins once the shape of the object has been drawn with thin lines in all the projection planes. In this phase, thicken the visible edges with a soft pencil of appropriate width. High precision is paramount.

First, draw the circles and fillets, followed by other lines. Begin with horizontal lines from the top to the bottom in all the projections and continue with all the vertical lines from left to right in all projections, finishing with all the oblique lines. During the process, correct all the minor errors in terms of matching and parallelism. There is no need to erase the thin lines in the corners.

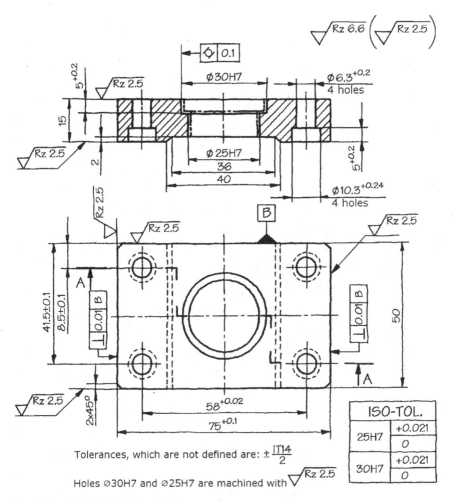

Fig. 2.38 Ninth sketching phase [10]

Sixth sketching phase (Fig. 2.35) follows once the visible edges have been thickened. In this phase, draw the invisible edges when they exist and continue by drawing the cross-section lines with a pencil of medium hardness and suitable thickness. This is followed by finishing all the other details of the object shape and marking, in order to achieve the required distinction between the importance of the individual lines, i.e., using lines of at least two thicknesses.

Seventh sketching phase (Fig. 2.36) is about adding blind measuring lines. Dimensioning rules prescribe the insertion of blind measuring lines, i.e., dimensions without numbers and symbols, in the order required by the manufacturing personnel for each working phase. Draw the supporting dimensioning and measuring lines

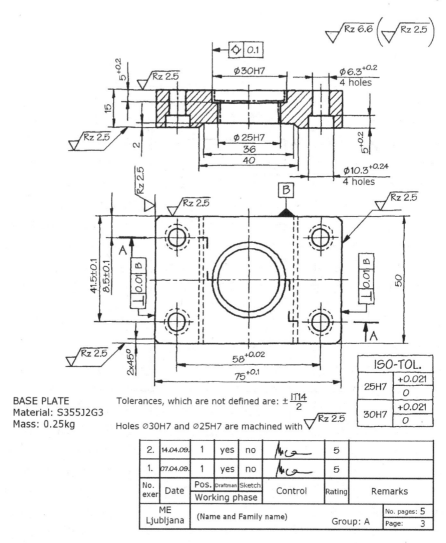

Fig. 2.39 Tenth sketching phase [10]

with a pencil of medium hardness and suitable thickness, and the arrows, with a soft pencil.

Eighth sketching phase (Fig. 2.37) follows once all the blind dimensions have been set and properly distributed. This is followed by measuring the object and inserting the dimensions. Real numbers are rarely inserted as the numbers are usually rounded off to whole numbers, to 0 or 5 or just 0. The angles must be checked during this phase. Next to all the angular dimensions insert the necessary special characters, such as the circular shape sign Ø.

Table 2.1 Activities for individual sketching phases

Phase no.	Procedural sketching (PS)	Stroke sketching (SS)	Notes
1	Position of objects, projections, cross-sections, dividing planes	Defining the position of objects, drawing criteria, main and supporting bisectors of the object	Phases 1, 2, 3 of PS, combined in SS
2	Spatial definition of a drawing, scale unit, position of views	Drawing visible and invisible edges with thin lines	
3	Drawing the main and supporting bisector in views	Drawing special shapes in the order: small circles, large circles, small arcs, large arcs and irregular curves	
4	Drawing the vital contours of the object in all views	Thickening visible and invisible edges to make the object recognizable	
5	Thickening and intensifying all contours, beginning with circles, followed by lines. Thin lines are not erased. All visible edges are drawn	Dimensioning, writing tolerances, processing, special instructions, drawing the main image, inserting notes for technological procedures, final form of the drawing	
6	Drawing invisible edges to improve the imaginability of the object		
7	Drawing all the dimension lines without values (drawing the values follows the sequence of technological operations and procedures)		
8	Inserting dimensions for dimension lines, dimensioning the object		
9	Inserting tolerances, special instructions and processing, hatching		
10	Equipping the drawing with the header, other marks, edges, etc.		

Ninth sketching phase (Fig. 2.38) includes inserting the units of length and shape, the position tolerances, the surface roughness, the marking and denoting the cross-section planes, inserting the necessary instructions and all the other data. This is followed by hatching of the cross-section areas, paying attention to not drawing over the numbers and the text on the drawing.

Tenth sketching phase (Fig. 2.39) is the last phase, i.e., the phase where the freehand sketching is completely finished. This phase includes filling in the parts list, the drafter's name, and—in the case of an assembly—also the position number within the assembly drawing.

2.4.2 Stroke Sketching

Some authors [20] suggest that it is possible to merge these ten phases of sketching 2D drawings into just five phases. The individual phases are defined in Table 2.1 and combine the individual phases of the ten-phase procedure, especially the first part. Stroke sketching is recommended for experienced designers.

Practising is the most important part of sketching. Courses that include sketching exercises have different names. Their common names are sketching, technical drawing, modelling, space modelling, etc.

In the case of computer space modelling, the working method with a pre-printed table is particularly recommended. This is a reliable way of introducing a beginner to the development process, from product abstraction to a digitised product model.

2.5 Sketching Spatial Drawings

The easiest way to present the shape of an object is with a spatial drawing, as this makes it possible to also present complex details or knots. In Europe, this method is very rarely used, while in the USA, it is very common. In all cases of spatial views, generally originating in an orthogonal projection, it is first necessary to determine a coordinate system and the direction of the main width and height axes. For spatial sketching, an isometric projection is the one most frequently used. In isometry it is very easy to present cylindrical objects or the details of a cylindrical shape. With cylindrical shapes being the most common ones in mechanical engineering, isometry has emerged as the most suitable way. But before deciding for a spatial view, you have to first check all the surfaces, define their dimensions and then draw them according to the coordinate system.

A course of sketching free-form surfaces is prescribed and depends on the shape of the object. One course is suitable for rotational parts, while the other is more suitable for parts made according to the principle of free-form surfaces. These parts and products include intermediate products, castings or pressed parts.

Let us first take a look at the process of sketching cylindrical, rotational objects. With rotational objects (see the example in Fig. 2.41) and freehand sketching, follow the working phases below [35]:

First sketching phase Draw the object's main centreline, define the distances and draw the conjugate diameters, according to the coordinate system. Continue by specifying the size of the end surfaces' conjugate diameters, relative to the actual

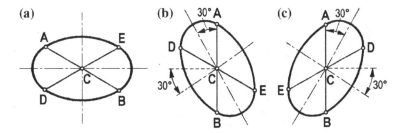

Fig. 2.40 Sketching the isometry of a rotational object. Two conjugate diameters defined centre lines for ellipse (**a**), isometric view of ellipse (**b**), opposite circle by isometric view (**c**)

dimensions, and the drawing scale. It should be pointed out that isometric sketching does not strictly follow the use of standard scales.

Second sketching phase Using thin lines, begin by designing the final ellipses. Usually, not all the lines that are required for a design are drawn. The simplified design approach is applied. In terms of designing, bear in mind that the centres of ellipse arcs are in the upper plane, on the horizontal and vertical axes. The side ellipses are designed by inclining their centrelines by 30° to the left and right, respectively, relative to the vertical line (Fig. 2.40b and c).

Third sketching phase Draw the remaining conjugate diameters at suitable distances according to the drawing scale. Using a thin line, draw concentric and other ellipses.

Fourth sketching phase Using thin lines, finish the presentation of the object. Connect the ellipses, draw slots, bores, fillets, ribs, nuts, threads and roughly specify the depths and all other details. In this sketching phase you often need to specify the visible and invisible parts of the object.

Fifth sketching phase Thicken the visible parts of the object. Using a dotted line, mark the invisible parts of the object (invisible edges) that significantly contribute to a clear view of the object in all its details. It should be noted that faithfully abiding by sketching rules will result in the thin lines not obscuring the clarity of the spatial view. By not following the sketching principles and drawing lines with the wrong thicknesses means you will have to erase the thin lines as they will not be easily distinguished from the heavy, dark ones. It is typical of freehand sketches that invisible edges are very rarely shown.

2.5.1 Presenting a Half Cross-Section

When for clarity reasons you want to show the inside of an object, cut out a quarter (1/4) of the object along its length x and width y (Fig. 2.42a), or half (1/2) of the object along the length x (Fig. 2.42b). When round surfaces are not in the direction of the object's dimensions, determine their centrelines and the approximate positions of

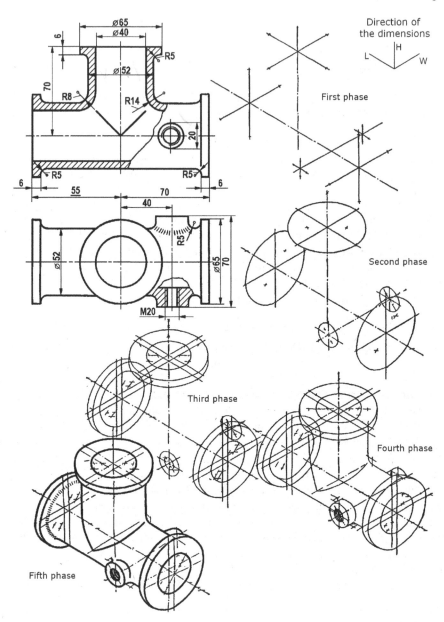

Fig. 2.41 Sketching the isometry of a rotational object [10]

the surfaces, relative to the part of the angle in reality, by dividing the corresponding angle in the isometry in the same direction.

Non-rotational objects (Fig. 2.43) are sketched in the following sequence:

(a)   ¼ cross section   **(b)**   ½ cross section

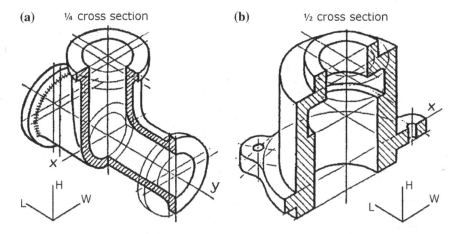

Fig. 2.42 Quarter (**a**) and half (**b**) cross-section in spatial view—isometry [10]

First sketching phase Using thin lines, draw the basic, unrounded shape of the object (rough shape of the object).

Second sketching phase Insert conjugate bore diameters and draw fillets, matching real dimensions.

Third sketching phase Design ellipses or parts of ellipses, if necessary.

Fourth sketching phase Using thin lines, finish the shape of the object by drawing the side parts (cut-outs, cut-ins, ribs, concentric ellipses, visible due to wall thickness, etc.).

Fifth sketching phase Thicken all visible edges. Using a dotted line, also thicken the invisible edges that contribute significantly to the image of the product. It should be noted that spatial views only show invisible edges in exceptional cases, and this is the key difference between 2D and spatial views.

Drawing in isometry is difficult for beginners, who get lost in the sheer number of lines and do not follow closely the dimension plan. For this reason, it is recommended to do exercises in drawing those elements that appear more frequently as structural elements in sketching.

Repeat the exercises by sketching these elements in different positions and views. So, it is recommended to do exercises in drawing frequently occurring elements (except round surfaces) in different positions. Referring to Fig. 2.44, they include [24, 35]:

(a) When drawing a hex nut in different positions in isometry, the following is important: the diameter circumscribing the nut, the spanner opening and the parallelity of the nut surfaces with object dimensions. In the direction of one dimension are the tops of the edges on the ellipsis, representing the nut-circumscribing diameter. In the direction of the other dimension is the spanner opening. The nut edges are parallel to the first dimension, i.e., the joining line of the tops. Using the conjugate nut-circumscribing diameters, design an ellipsis, specify the tops of a

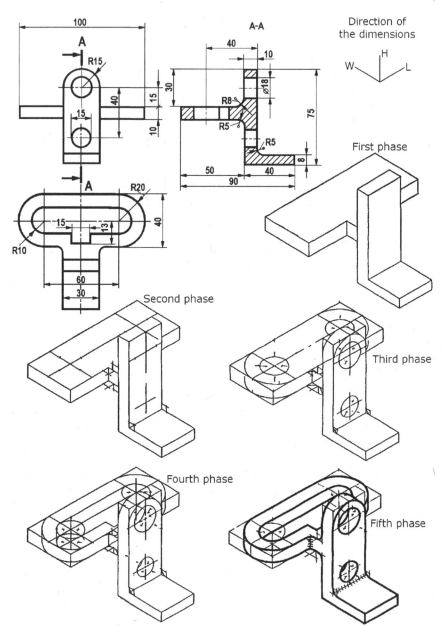

Fig. 2.43 Sketching the isometry of a non-rotational object [10]

hexagon, insert the spanner opening *s* onto the other dimension and specify the hexagon. Continue by specifying the thickness (height) of the nut in the direction of the third dimension and the parallel drawn surface. Drawing nut chamfering is approximate. Follow a similar procedure to draw an octagonal nut.

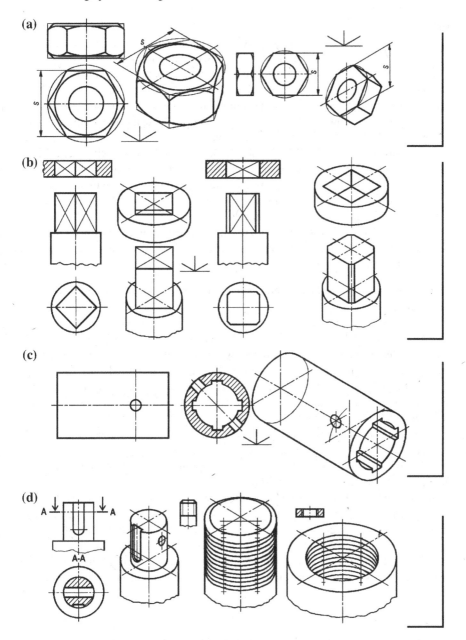

Fig. 2.44 Isometric view of the most frequently used elements (**a**—hex nut, **b**—four-sided extensions and holes, **c**— slots, bores and **d**—threads for nuts and bolts) [24]

(b) In isometry, four-sided extensions and deepenings are rotated from a rectangular projection. If there are two rectangular solid surfaces visible in a rectangular projection, there is only one visible in isometry and vice-versa. This originates in a proper observation of the dimensions and the proper use of the dimension plan, as shown in Fig. 2.43.

(c) In isometry, slots and bores are placed relative to the axis position; not in the same positions, however, but by sight, like in a rectangular projection. It is vital to have a good vision of which dimensions of such parts are visible and which dimensions they are parallel to. They are inserted in isometry with their actual distances and sizes, relative to the orientation of the main dimensions.

(d) Drawing thread for an explicit view in isometry follows the approximation procedure. Each groove is shown with an equidistant ellipse,[2] and the space between the ellipses matches the thread pitch (or approximately if the pitch is small). Because all these ellipses have identical radii of curvature, it is necessary to draw parallel lines into all three centres of the ellipse arcs, insert distances and draw the arcs one by one.

We have presented all the elements that are important for an engineering view of the environment.

It should be noted that sketching is not just a matter of engineering but also serves as a basis for other human expressions, such as art. The message of an artistic sketch is based on the metaphysics of the artist's cognition of the world, so an artistic sketch conveys a different message. In both cases, proficiency is represented by the creativity of a creator, so the expressing exclusiveness of the respective creations (engineering and artistic sketches) is unique to each creator. It also shows how important an engineer's sketch is.

It is possible to transfer models from the analogue into the digital world, managed by a computer. When it is about a complete relocation of a technical system in the real world, pure scanning, this transition is possible. Namely, dimensionally or proportionally the same space is used [7].

In practice, engineers' inferior skills are recognizable by many models of technical systems being developed without a sense of proportionality. During the last decade (up to 2009), a good example of this argument is often seen in an imperfect understanding of the space inside and under the bonnet, in the headlight area. Such a treatment of space and not understanding proportionality often results in difficult bulb replacements in some headlights.

---

[2] Equidistant ellipse.

Chapter 3
3D Modelling

Abstract Space modelling uses geometric topological elements. The chapter presents all the basic topological elements, as well as the matrix calculus for translation, rotation, scaling, mirroring and perspective. These are the key operations for space modelling.

In order to represent objects or products in a virtual environment, many types of equipment have been used since early times. Besides analogue records in 3D space (such as a sketch, an illustration or a photograph), solutions for the best possible form of digital recording were investigated. When the computer processor and the corresponding mathematical relations appeared they offered, for the first time, an opportunity for a high-quality digital record of 3D space. Of course, this had to be accompanied by appropriate 3D modelling methods, and several procedures and approaches appeared during their development. The approaches were, however, initially very different. After some development, some methods established themselves and proved to be suitable for computer processing. Seeking the right solutions was necessary and important, particularly from the viewpoint of getting a high-quality basis for digital recording. However, the issue of standardizing communications between the user and the computer has remained open.

Historically, development progressed gradually, from simply describing objects with wireframe models and a surface description of 3D models to the solid model, as the most reliable way of describing real models in space (Fig. 3.1).

Increasing computer capacities gradually provided users with new solutions based on different modellers. However, all modeller developers need to follow the global user. Standard imaging has been established in global engineering practice, i.e., it is understood by engineers and technicians in different parts of the world. This was the reason why the large number of original software solutions was gradually reduced to a smaller number of providers for general 3D-modelling equipment. However, there are also developers for specific needs and forms on the market.

To understand the structure of a 3D model it is vital to also understand the fundamentals of the topological definition of a geometric model. Below, the basic topological elements will be presented: point, edge, loop, surface, and volume. This is

J. Duhovnik et al., *Space Modeling with SolidWorks and NX*,   49
DOI: 10.1007/978-3-319-03862-9_3, © Springer International Publishing Switzerland 2015

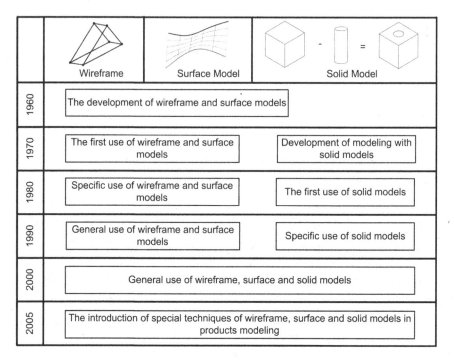

Fig. 3.1 Development of 3D modelling

followed by presenting the different types of geometric models, i.e., from a wire-frame and surface model to a solid model. The result is geometrical transformations that allow the representation of an object on the screen and its manipulation.

3.1 Topological Elements in a 3D Modeller

The topology of an object in space consists of five topological elements, with each of them having its own characteristics. The geometric modeller's database considers the method of describing the basic topological elements and their relations.

The basic topological elements include: point, edge, loop, surface and volume, all of which can be used to represent a 3D model of an object (Fig. 3.2). The parameters of each of the described topological elements and their characteristics are presented.

Any object, however simple, represents a volume in nature. This volume is described by the surface, dividing the object's interior from its environment. Each surface is surrounded by a loop, consisting of a finite set of edges. Each edge is defined by at least two points in space, representing its beginning and end.

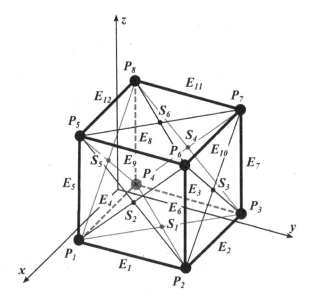

Fig. 3.2 Topological elements on a model

3.1.1 Point

The point is the basic topological element of 3D space; its position in space is described by coordinates that depend on the chosen coordinate system.

3.1.1.1 Cartesian Coordinate System

$$P = P(x, y, z)$$

A point in a Cartesian coordinate system is specified by its distance from the main three planes: the x coordinate—distance from the front view y–z plane; the y coordinate—distance from the side view x–z plane, the coordinate z—distance from the top view x–y plane (Fig. 3.3).

3.1.1.2 Cylindrical Coordinate System

$$P = P(r, \varphi, z)$$

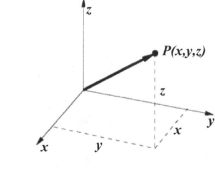

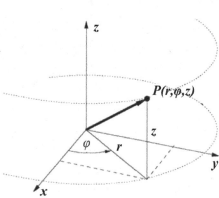

This is an upgrade of the so-called polar coordinate system, where point T is determined by a distance from the origin, i.e., the local vector (r), the rotation of the
point around the z axis and the axis itself by the revolution angle (φ) relative to the
x–z plane. The (z) coordinate is added, representing movement in the z direction
(Fig. 3.4).

   A cylindrical coordinate system is used for different types of rotated or turned
parts. It is useful for a computer-controlled lathe. Figure 3.4 presents the (r) radius,
acquired by moving the knife, mounted on the lathe support, the revolution angle
(φ), the (z) coordinate represents movement along the lathe's bed.

3.1.1.3 Spherical Coordinate System

$$P = P\,(r, \varphi, \theta)$$

In contrast to a cylindrical coordinate system, where the height is specified by the z
coordinate, the height in a spherical coordinate system is specified by the revolution
angle θ between the point's local vector and the x-y plane (top view). The distance
from the centre is specified by the radius of a sphere (Fig. 3.5).

Fig. 3.5 A point in a spherical coordinate system

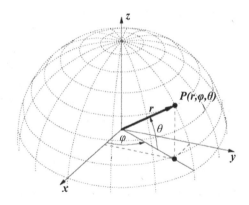

A spherical coordinate system is used for all products that are either built or operate in the shape of a sphere. The most well-known application of a spherical coordinate system is Gauss-Krueger's coordinates for determining the positions on a globe. These coordinates are used by surveyors, who determine two types of coordinate: the latitude (φ) and the longitude (θ). A more accurate determination would generally also require a distance from the centre of the Earth (r); however, it varies from place to place in different parts of the world, and instead we use the height above sea level, according to conventional bases in different parts of the world. It is to be expected that some definitions of the radius itself and sea level will change again.

3.1.2 Edge

$$E \Rightarrow f(x, y, z)$$

The edge is a topological element—the connection between at least two points (Fig. 3.6). In a linear connection, the edge is represented by the line, defined by its beginning (P_1) and end (P_2) points.

$$E \Rightarrow (P_1, P_2)$$

The edge can be generally presented as a function (curve) between two points that are running in space through a defined point. The function can be of any order. As a rule, it should be as close as possible to the function of a natural phenomenon, taking place on its surface or in its vicinity and defined by the said function. For example, hydraulic phenomena (vessels at sea, the blade of a Kaplan turbine, etc.) or aero phenomena (windmills, cars, car spoilers etc.) are described by the Bernoulli equation, which makes it possible for a surface to be designed by a fourth-order curve, and not at all by a third-order curve.

Fig. 3.6 Possible connections
between two points with
different functions in space

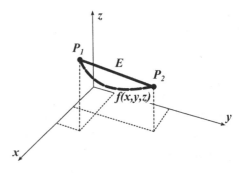

Fig. 3.7 A closed loop in
space

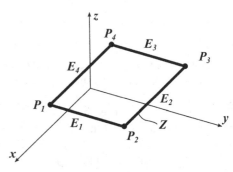

3.1.3 Loop

$$L \Rightarrow (E_1, E_2, ..., E_n)$$

The loop consists of a set of edges, connected one to another in one or another way
(Fig. 3.7).

A closed loop represents sequentially connected and closed edges. The said conditions provide logically executed and repeatable computer operations. An open loop
is referred to as an edge assembly or polylines. The polylines start by defining the
first point, which is then in the following step defined as the last point with n data
polylines. The number n specifies the number of connecting lines or curves of particular polylines. Such designed polylines can be quickly transformed into different
types of curves. The method is very useful for specifying the regression lines or
curves with an order m.

An open loop represents a set of edges where the last edge with its second point
does not connect with the first point of the first edge. An open loop can appear with
wireframe models. Surface or volume models cannot use an open loop, and surfaces
in open loops cannot be defined.

Fig. 3.8 A topological
element—surface

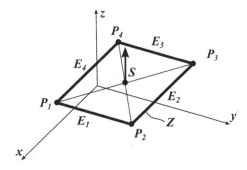

More frequent and useful for modelling purposes is the so-called closed loop with
a number of edges finishing in the initial point. Only a closed loop, lying on a plane,
can represent a surface.

3.1.4 Surface

$$S \Rightarrow L \Rightarrow (E_1, E_2, ..., E_n)$$

The surface is represented by a closed loop. A surface is defined by a loop and a
normal vector to a surface. A surface is required to manage surface and solid models
(Fig. 3.8). Wireframe models can be presented without defined surfaces. For simple
surface-model presentations, there are simple laws used to create the surface filling.
Surfaces sometimes come with surface colour information or even a "pattern" to
improve the image of the object itself.

Models that are presented with surfaces only, are very useful for simulating phenomena in the environment. However, when reshaping the model, their major drawback is errors in the surface connection, resulting in inaccurate details.

3.1.5 Volume

$$V_1 \Rightarrow (S_1, S_2, ..., S_n)$$

The volume is the final element in a geometric modeller. The volume is described as
a set of surfaces, dividing the object's interior from the exterior—the environment.
The volume is characterized by the requirement that all surfaces need to be connected
and all normals should point outwards (Fig. 3.9).

Fig. 3.9 Volume as a complex
topological element

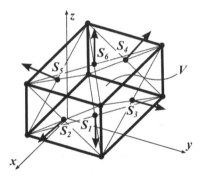

Similar to surfaces, where a loop may not have any undefined edges, the volume also allows no undefined surfaces. Volume elements are required to model solid bodies.

Modellers were also developed in line with the topological structure. Looking at the required parameters for a single topological element, one can see that at the very beginning of modeller development, memory was the main obstacle. With the arrival of sufficiently capable RAM units (exceeding 4 Gb), modeller development accelerated in the direction of digitising objects.

This all suggests that the development of useful methods and modelling largely depends on the development of other systems, which is computer power in this case.

3.2 Presenting 3D Models

3.2.1 Wireframe Model

The first 2D and 3D modelling computer programmes were based on two topological elements: the point and the edge. With the development of 3D modelling, other topological elements came into use.

The wireframe model is the simplest way to describe an object in 3D space. Representing a model, its edges are shown as lines, connecting the points in space with a wire.

Containing only vertices and edges and their relations, the databases for wire-frame modellers are simple. Due to simple relations and a small volume of data, the computation is fast and does not require a lot of computer memory.

A drawback of wireframe modellers is that an image with a larger number of edges becomes unclear. The clear orientation of an object in space can be lost. The incorrect connections of edges between individual points can lead to an anomaly, as shown in Fig. 3.10.

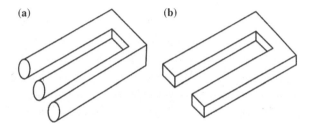

Fig. 3.10 Anomaly of a wireframe model, caused by incorrect connections between points: incorrect (**a**) and correct (**b**) connections

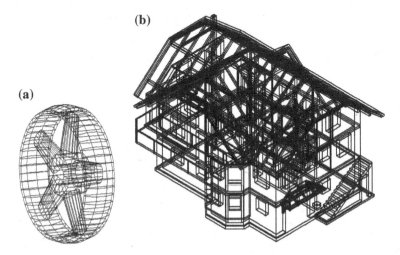

Fig. 3.11 Examples of a wireframe models: a pulley (**a**) and a house (**b**)

Wireframe models are not suitable for computer-aided analyses (CAA) as they do not allow the calculation of even basic quantities, such as the model's surface or volume.

3.2.2 Surface Model

The surface model is an upgrade of the wireframe model. For the wireframe model, the loop is the highest topological model. The surface model requires all loops with defined edges, plus normal vectors for each closed loop. In terms of the surface, the models are divided according to the method of describing the surface. The following description methods are the most commonly used:

- surfaces as parts of the shape of geometric bodies (Fig. 3.12),
- interpolate surfaces (Fig. 3.13),

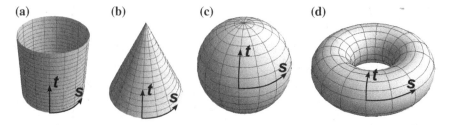

Fig. 3.12 Surface as part of the shape of a cylinder (a), a cone (b), a sphere (c) and a torus (d)

Fig. 3.13 An interpolate
surface, defined by two edge
curves

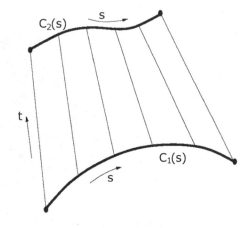

- parametric surfaces (Fig. 3.14),
- polygon meshes (Fig. 3.15).

3.2.3 Volume (Solid) Model

Solid models can be defined by a Euclidean space, defined by two regions, i.e., the
interior one and the exterior one, divided by the boundary of the object. Its boundary
is defined by at least one closed surface and/or a set of connected open surfaces. A
common feature of all the representations of solid models is that the interior of the
object consists of a number of points, geometrically closed by the boundary of the
object.

3.2.3.1 Instances and Parameterized Shapes

This method describes simple and similar shapes of objects with the use of basic,
i.e., parameterized, shapes. Figure 3.16 shows a couple of examples of new shapes,

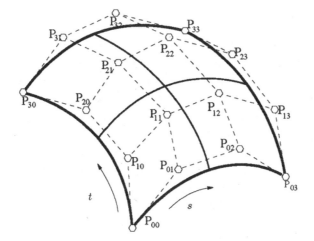

Fig. 3.14 A free parametric surface, described with 16 control vertices

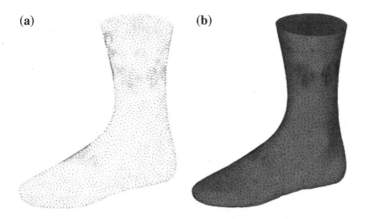

Fig. 3.15 A polygon mesh of a human foot: a scanned surface (**a**) and a triangle mesh (**b**)

designed by simple linear transformations of existing models, such as a unit sphere, a cube and a cylinder. A family of similar shapes can be created by parameterizing instances. All the resulting variants can be created by changing the parameters. Figure 3.17 shows an example of an instance and the parameterization of a Z-profile.

3.2.3.2 Boundary Representation

A boundary representation (B-Rep) is based on an argument that a physical object is closed from all sides with boundary faces (surfaces), dividing the model from the rest of the space. Each face is limited by edges and they are limited by vertices (points). Figure 3.18 shows a boundary-represented object.

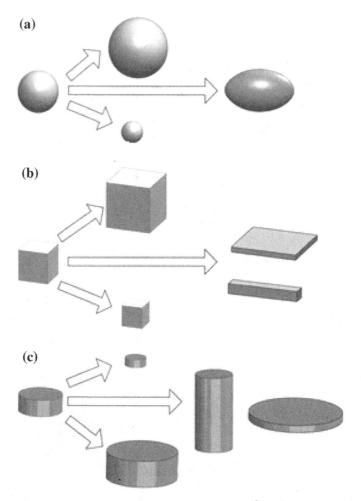

Fig. 3.16 Instances and linear transformations or better parameters extrude: a sphere (**a**), a cube (**b**) and a cylinder (**c**)

3.2.3.3 Constructive Solid Geometry

The constructive solid geometry (CSG) method represents one of the most popular techniques for generating solid models. The method is simple—for both understanding and communicating with the user. A 3D model validation according to this method is simple.

The method is based on the principle that any physical body can be generated as a combination of elementary shapes, i.e., primitives. A large set of primitives can be used. In practice, the most frequently used ones include a rectangular solid, a cylinder, a cone and a sphere, as shown in Fig. 3.19.

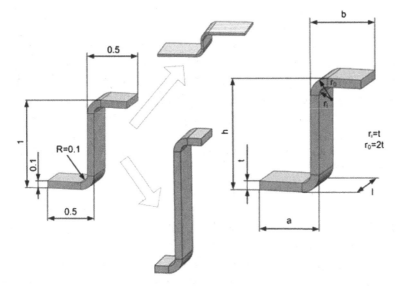

Fig. 3.17 Parameterization of a Z-profile

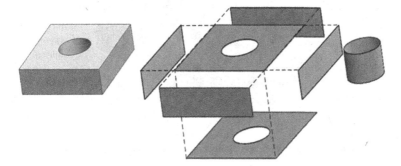

Fig. 3.18 An object with boundary surfaces

3.2.3.4 Feature-Based Mdelling

Feature-based modelling is a 3D modelling technique that makes it possible to build a model at a higher level, such as manipulation with the basic geometric entities (point, line, etc.) or primitives. A model is represented as a combination of the CSG and B-rep algorithms. Besides the basic geometric and the topological modelling structure, the method also takes account of higher information levels, such as the geometric characteristics of holes, slots, fillets and other shapes.

A feature represents a set of geometric entities that appear and are recorded in a certain order. It allows—by means of a couple of simple operations—the creation of a large number of geometric primitives that make up individual model parts (Fig. 3.20). The features are very useful for the user, as they allow him or her to upgrade a feature

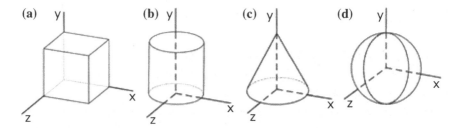

Fig. 3.19 Examples of the basic elementary shapes, i.e., primitives: a cube (**a**), a cylinder (**b**), a cone (**c**) and a sphere (**d**)

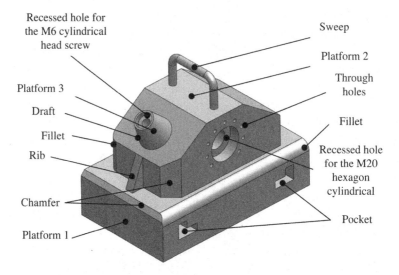

Fig. 3.20 Using features for modelling a product of moderate complexity

that represents a certain manufacturing technology or its form. The features bring the user close to technological manufacturing operations, which significantly improves the practical value of a modeller.

The basis for feature manipulation (Fig. 3.21) are Boolean operations:

- union,
- difference,
- intersection.

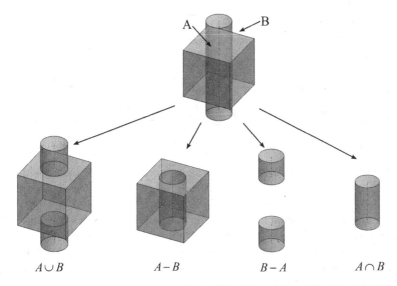

Fig. 3.21 Boolean operations between the objects (from *left* to *right*): union ($A \cup B$), difference ($A - B$ and $B - A$) and intersection ($A \cap B$)

3.3 Geometric Transformations

A 3D graphics user would like to observe the scene from different points of view and move some objects in space relative to other objects. These operations are made possible by **geometric transformations**. They are used for the purpose of positioning, orienting and scaling the objects, as well as for mirroring, perspective view, etc. Before describing the transformations, let us take a look at some mathematical operations that make the passage into matrix formulations possible, as these are the easiest way to perform transformations.

When mirroring objects from the real world into the virtual one, it is vital to ensure independent imaging, independent of the size and type of display. In order to achieve that, a new environment, a new space—called a uniform space—has to be created. A uniform space should provide complete neutrality for different types of mirroring from the real into the virtual display world (Fig. 3.22).

Homogenous coordinates are used in a uniform space. A 3D description of a point is translated into homogenous coordinates by adding to Cartesian coordinates $\{x, y, z\}$ a fourth component w, also called a homogenous coordinate.

$$\mathbf{p} = \left\{ \begin{array}{c} x \\ y \\ z \end{array} \right\}, \quad \mathbf{p}^H = \left\{ \begin{array}{c} x \\ y \\ z \\ w \end{array} \right\}$$

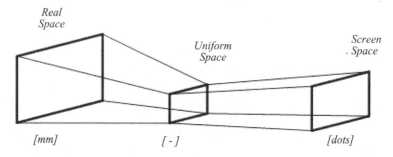

Fig. 3.22 Mirroring from the real into the virtual display world by means of a uniform space

An inverse transformation that projects homogenous coordinates back into a 3D space is called a *projection*. The neutrality to newly introduced coordinates w is established by its value 1. All the values from the real world are therefore projected onto a display of a certain resolution by practically retaining—in a homogenous space—identical values, identical dimensions.

3.3.1 Translation

With the geometric transformation that is to be used to move point **p** to **p′**, the translation is performed by specifying the size of the movement by the vector $\mathbf{t} = \{T_x, T_y, T_z\}$. In a matrix equation and using homogenous coordinates, the transformation can be formulated as follows (Fig. 3.23):

$$\mathbf{p}' = \begin{Bmatrix} x + T_x \\ y + T_y \\ z + T_z \\ 1 \end{Bmatrix} = \begin{Bmatrix} 1 \cdot x + 0 \cdot y + 0 \cdot z + T_x \cdot 1 \\ 0 \cdot x + 1 \cdot y + 0 \cdot z + T_y \cdot 1 \\ 0 \cdot x + 0 \cdot y + 1 \cdot z + T_z \cdot 1 \\ 0 \cdot x + 0 \cdot y + 0 \cdot z + 1 \cdot 1 \end{Bmatrix} = \begin{bmatrix} 1 & 0 & 0 & T_x \\ 0 & 1 & 0 & T_y \\ 0 & 0 & 1 & T_z \\ 0 & 0 & 0 & 1 \end{bmatrix} \cdot \begin{Bmatrix} x \\ y \\ z \\ 1 \end{Bmatrix}$$

This equation can also be written in a more concise way:

$$\mathbf{p}' = \mathbf{T} \cdot \mathbf{p}$$

where **T** is the transformation matrix:

$$\mathbf{T} = \begin{bmatrix} 1 & 0 & 0 & T_x \\ 0 & 1 & 0 & T_y \\ 0 & 0 & 1 & T_z \\ 0 & 0 & 0 & 1 \end{bmatrix}$$

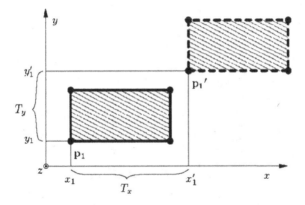

Fig. 3.23 An example of translating an object on a plane

The matrix yields more accurate values if only translations are performed. But when translations are performed together with rotation, especially with large translation values, it can result in large discrepancies due to the increased influence of the asymmetric values of individual matrix elements. For this reason, translations are generally performed independently and not together with other transformations.

3.3.2 Rotation

A rotation transformation rotates a selected point **p** into a point **p**$'$ about one of the coordinate axes by the revolution angle ϕ. Figure 3.24 shows an example of rotating a rectangle. When rotating about the coordinate axis z, the coordinates of the rotated point are calculated as follows.

With matrix calculus and using homogenous coordinates, this transformation can be presented as:

$$\mathbf{p}' = \mathbf{R_z} \cdot \mathbf{p}$$

where $\mathbf{R_z}$ is a transformation matrix, rotating a chosen point about the z axis:

$$\mathbf{R_z} = \begin{bmatrix} \cos(\phi) & -\sin(\phi) & 0 & 0 \\ \sin(\phi) & \cos(\phi) & 0 & 0 \\ 0 & 0 & 1 & 0 \\ 0 & 0 & 0 & 1 \end{bmatrix}$$

Analogous to this is the remaining two rotations in space, i.e., rotations about the y and x axes:

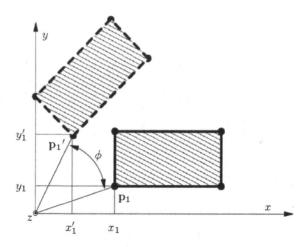

Fig. 3.24 Rotating an object about the z-axis

$$
\mathbf{R_y} = \begin{bmatrix} \cos(\phi) & 0 & -\sin(\phi) & 0 \\ 0 & 1 & 0 & 0 \\ \sin(\phi) & 0 & \cos(\phi) & 0 \\ 0 & 0 & 0 & 1 \end{bmatrix}
\qquad
\mathbf{R_x} = \begin{bmatrix} 1 & 0 & 0 & 0 \\ 0 & \cos(\phi) & -\sin(\phi) & 0 \\ 0 & \sin(\phi) & \cos(\phi) & 0 \\ 0 & 0 & 0 & 1 \end{bmatrix}
$$

As explained later on, each of these three rotations can be combined in any order into a general 3D rotation. It should be noted that great differences can appear in the ranges between 0° and 5°, and 85° and 90°. The resulting differences in the numerical part can be corrected with two operations: rotation is performed by first executing it in a negative direction and an angle of 45 − alpha/2 is deduced, followed by rotation in a positive direction by an angle of 45 + alpha/2, where the angle alpha represents the required rotation, set by the user. The decision for such an operation should be made when the alpha angle is smaller than a specified value. More advanced modellers have this method built into their programme code, while the more basic ones do not include it. As a result, rotations of less than 5°, for example, are becoming unreliable after several repetitions, and distorted objects and ratios between the surfaces begin to appear.

3.3.3 Scaling

A general form of a scaling matrix, defining the scaling matrix elements, should first specify the size of the scaling. The size is specified according to the origin of the coordinate system and provides different enlarging or shrinking factors in a chosen direction. To scale the point $\mathbf{p} = \{x, y, z\}^T$ by the chosen scaling factors S_x, S_y, S_z the following form of matrix calculus can be used:

Fig. 3.25 Scaling an object

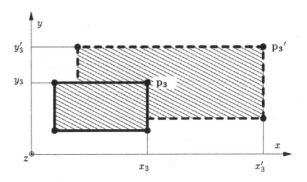

$$\mathbf{p}' = \mathbf{S} \cdot \mathbf{p}$$

whereby **S** is a scaling matrix:

$$\mathbf{S} = \begin{bmatrix} S_x & 0 & 0 & 0 \\ 0 & S_y & 0 & 0 \\ 0 & 0 & S_z & 0 \\ 0 & 0 & 0 & 1 \end{bmatrix}$$

When all three factors are identical this is called uniform scaling. When the scale factor S_i is larger than 1, this is enlarging in a given direction, and when S_i is smaller than 1 and larger than 0, this is an object contraction. Figure 3.25 shows an example of non-uniform rectangle scaling, i.e., when the S_x, S_y, S_z scalars are different for each coordinate.

A scaling matrix has only diagonal elements, which makes it robust and reliable also for larger scaling values, using the **ZOOM** function, for example. It is a known fact that the accuracy of the details can be maintained without problems for enlargements by a factor up to 400.

3.3.4 Mirroring

Mirroring is a special form of scaling, where some of the S_i factors are identical to -1. When $S_x = -1$, this is mirroring across the yz plane. When $S_y = -1$, this is mirroring across the xz plane, and finally, when $S_z = -1$, this is mirroring across the xy plane.

$$\mathbf{p}' = \mathbf{Z} \cdot \mathbf{p}$$

whereby:

• mirroring across the y-z plane (Fig. 3.26)

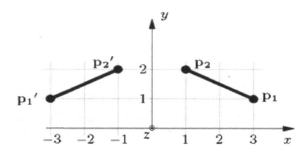

Fig. 3.26 An example of mirroring a line across the y–z plane

$$Z_{yz} = \begin{bmatrix} -1 & 0 & 0 & 0 \\ 0 & 1 & 0 & 0 \\ 0 & 0 & 1 & 0 \\ 0 & 0 & 0 & 1 \end{bmatrix}$$

• mirroring across the x-z plane

$$Z_{xz} = \begin{bmatrix} 1 & 0 & 0 & 0 \\ 0 & -1 & 0 & 0 \\ 0 & 0 & 1 & 0 \\ 0 & 0 & 0 & 1 \end{bmatrix}$$

• mirroring across the x-y plane

$$Z_{xy} = \begin{bmatrix} 1 & 0 & 0 & 0 \\ 0 & 1 & 0 & 0 \\ 0 & 0 & -1 & 0 \\ 0 & 0 & 0 & 1 \end{bmatrix}$$

3.3.5 Perspective Projection

A perspective projection is required to show the depth of an object. The objects are deformed by showing the closer objects larger than the more distant ones. The perspective transforms the parallel lines into lines that converge to the vanishing point. In terms of the number of vanishing points, there are (a) one-point, (b) two-point and (c) three-point perspective projections (Fig. 3.27).

General presentation of the perspective projection in a matrix form:

$$\mathbf{p'} = \mathbf{W} \cdot \mathbf{p}$$

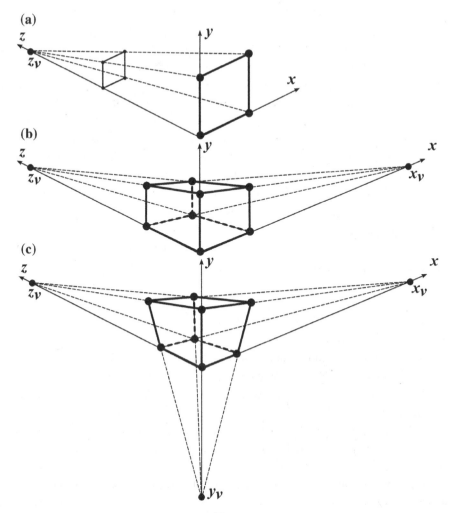

Fig. 3.27 Perspective projection. One-vanishing point (**a**), two-vanishing point (**b**), three-vanishing point (**c**)

where **W** is the transformation matrix for a perspective projection:

$$\mathbf{W} = \begin{bmatrix} 1 & 0 & 0 & 0 \\ 0 & 1 & 0 & 0 \\ 0 & 0 & 1 & 0 \\ p_x & p_y & p_z & 1 \end{bmatrix}, \text{ where: } p_x = -\frac{1}{x_v},\ p_y = -\frac{1}{y_v},\ p_z = -\frac{1}{z_v}$$

Let us take a look at projecting the point P onto the x-y ($z = 0$) plane, where the point of the observation is on the z axis (Fig. 3.28). Considering the similarity of triangles, we can write:

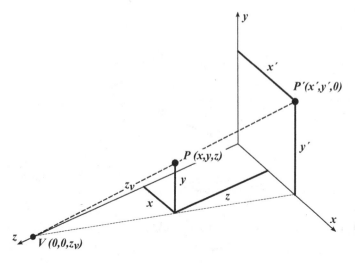

Fig. 3.28 Perspective projection of a point onto a plane ($z = 0$)

$$\frac{x'}{x} = \frac{x}{z_v - z} \Rightarrow x' = \frac{x}{1 - \frac{z}{z_v}} \quad \text{and} \quad \frac{y'}{y} = \frac{y}{z_v - z} \Rightarrow y' = \frac{y}{1 - \frac{z}{z_v}}$$

As this is a projection on the x-y plane, we know that $z = 0$. In this case, the perspective projection can be written as:

$$\mathbf{W}_y = \begin{bmatrix} 1 & 0 & 0 & 0 \\ 0 & 1 & 0 & 0 \\ 0 & 0 & 0 & 0 \\ 0 & 0 & p_z & 1 \end{bmatrix}, \text{ or } \begin{Bmatrix} x' \\ y' \\ z' \\ 1 \end{Bmatrix} = \begin{bmatrix} 1 & 0 & 0 & 0 \\ 0 & 1 & 0 & 0 \\ 0 & 0 & 0 & 0 \\ 0 & 0 & p_z & 1 \end{bmatrix} \begin{Bmatrix} x \\ y \\ z \\ 1 \end{Bmatrix} = \begin{Bmatrix} x \\ y \\ 0 \\ p_z \cdot z + 1 \end{Bmatrix}$$

Projecting the point $\mathbf{p}'$ onto a plane ($w = 1$) results in the point p' in three-dimensional space:

$$\begin{Bmatrix} x' \\ y' \\ z' \end{Bmatrix} = \begin{Bmatrix} \dfrac{x}{1 + p_z \cdot z} \\ \dfrac{y}{1 + p_z \cdot z} \\ 0 \end{Bmatrix}$$

Assuming that $p_z = -\frac{1}{z_v}$, this results in:

$$\begin{Bmatrix} x' \\ y' \\ z' \end{Bmatrix} = \begin{Bmatrix} \dfrac{x}{1 - \dfrac{1}{z_v} \cdot z} \\ \dfrac{y}{1 - \dfrac{1}{z_v} \cdot z} \\ 0 \end{Bmatrix}$$

Chapter 4
3D-Modelling Software Packages

Abstract This chapter presents space-modelling techniques with the use of features. The method is presented through two modelers—SolidWorks 2014 and Siemens NX 9.0. The structure of the menus and the main sub-menu commands are briefly described. The reader can gain a general insight into the commands and the structure for accessing individual commands, which is of great benefit for later use and for understanding the procedure. Such a structure will provide the reader with a comprehensive overview and understanding of the communications between the programme and the user.

4.1 Introduction

3D modelling provides us with a digitization of a facility with a precise definition of its location and orientation in relation to the selected coordinate system. If we connect the starting points and the orientations of the local coordinate systems with the introduction of a global coordinate system to the whole, we can determine every building within a global and universal place. It is possible to describe the geometry of buildings with the topology of their building blocks. For different orientations or a conversion we use a geometric transformation. If we are making any change, we write it from the basic to a temporary base. The operations are uniquely determined; therefore, almost the same procedures are used by all modellers. However, differences can occur from communications between the user and the software when using different modellers. Here, we are talking about a variety of communication techniques: some are more oriented towards computer-information communications, while others (the most recent ones) are based on general human or a clearer engineering communication. We believe that there will be time when it will be necessary to introduce high-quality standards for user-computer communications. In our case the user is represented by the engineer with a global engineering perspective on nature and resources.

J. Duhovnik et al., *Space Modeling with SolidWorks and NX*,   71
DOI: 10.1007/978-3-319-03862-9_4, © Springer International Publishing Switzerland 2015

This book presents two modellers that are different in terms of communications. Each has its own purpose and message for the user. The user should be able to recognize the suitability and whether to use one or another on different levels of use. We should emphasize that the communications technology remains as an insufficiently researched area, which means that it is hard to provide a solid judgement about the advantages of one or other approach.

4.2 SolidWorks

SolidWorks is a 3D modelling software package suitable for *Windows* operating systems. Due to its ease of use and performance it is becoming increasingly popular among users. Its strength is a simple communication language and a clear presentation of the steps in the planning as well as in the transformations. In the following chapters we will learn how to use this modeller. Later on, groups of typical commands will be described in detail in subsequent chapters and presented as discrete models. It is important that a user obtains an insight into groups of commands and their results first, after which he or she can apply the approach to direct use in modelling.

After starting the SolidWorks software (*Start* > *Programs* > *SolidWorks* > · · ·) we open a basic window. Let us go through the main parts of the menu that we will use later. Different software packages use different approaches: from the top left to the bottom right, or from the bottom left to the top right. The approach is comparable to the style of writing texts in different cultures and the user needs to understand it. For this reason it is preferable to examine each group of commands. When we create a general image of the distribution of the groups of menus on the screen, we can begin to search and select individual commands.

On the start screen, we first choose between creating a new document or opening an existing one. When creating a new document the software asks us for the type of document we want to design (Fig. 4.1). We can choose between *Part*, *Assembly* and *Drawing*.

The working environment opens after confirming the selected document. The general form of the working environment in the user interface is always the same, only individual commands that are specific to particular types of documents are changing. The user interface for modelling parts that is shown in Fig. 4.2 contains the seven main groups of widgets belonging to the software.

4.2.1 Menu Bar

The menu bar in *SolidWorks* (Fig. 4.3) consists of the most commonly used commands, the search field, assistance and the commands to manipulate the program window.

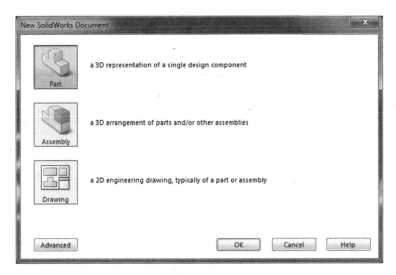

Fig. 4.1 The dialogue box for selecting a new document

It should be emphasized that the menus in each of the software versions are different, which is sometimes difficult to understand for the user. However, this can justified because of the search for a simpler and better communications language.

4.2.2 Command Manager and Toolbars

Command manager combines groups of commands that are specific to each operation in the product-modelling process. Figure 4.4 shows each of the command groups. By right-clicking on the *Command manager* and collecting particular groups we can activate/deactivate an individual command group.

4.2.3 Heads-Up View Toolbar

The *Heads-up view toolbar* (Fig. 4.5) enables us to choose between the different views in the modelling process. It ensures transparency within a chosen look and contains tools for manipulation with views. Other possibilities and commands can be viewed by clicking the icon with an arrow on the right. The reader is advised to have a detailed look at the commands and imagine using each of them. In this way the user will understand and consolidate individual commands in advance, so he or she will be able to use them quickly and easily while modelling. The commands are listed below in the order that they appear in the SW software version 2014.

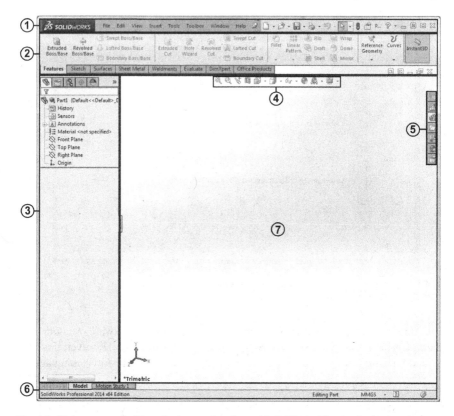

Fig. 4.2 Main groups of widgets in the user interface of *SolidWorks*. Number in the figure present the area for: Menu bar *1*, Command Manager and Toolbars *2*, Manager window *3*, Head-up view toolbar *4*, Task pane *5*, Status bar *6*, Graphic area *7*

4.2.4 Manager Window

The *Manager window* (Fig. 4.6) is located to the left of the graphic area. It contains **Feature Manager**, **Property Manager**, **Configuration Manager,** etc. (Table 4.1). We can select individual windows with a simple click on the tab at the top.

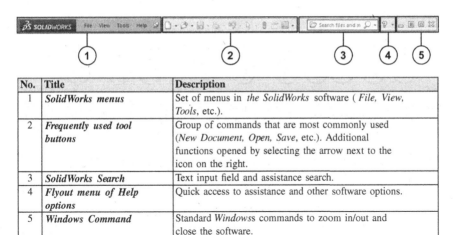

No.	Title	Description
1	*SolidWorks menus*	Set of menus in *the SolidWorks* software (*File, View, Tools*, etc.).
2	*Frequently used tool buttons*	Group of commands that are most commonly used (*New Document, Open, Save*, etc.). Additional functions opened by selecting the arrow next to the icon on the right.
3	*SolidWorks Search*	Text input field and assistance search.
4	*Flyout menu of Help options*	Quick access to assistance and other software options.
5	*Windows Command*	Standard *Windows*s commands to zoom in/out and close the software.

Fig. 4.3 *Menu bar* in *SolidWorks 2014* software

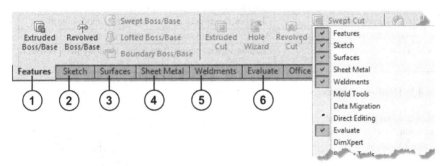

No.	Title	Description
1	*Features*	Command set for basic (extrusion, revolving, curve sweep, loft, etc.) and deduced (fillet, chamfering, patterning, mirroring, etc.) features.
2	*Sketch*	Command set for making sketches (individual entities, dimensions, relations, etc.).
3	*Surface*	Command set for surface modelling.
4	*Sheet Metal*	Command set for modelling sheet product.
5	*Weldments*	Command set for modelling welded constructions.
6	*Evaluate*	Command set for model constraints (dimensions, mass, tolerances, etc.)

Fig. 4.4 Command groups and window for activation/deactivation of the command groups

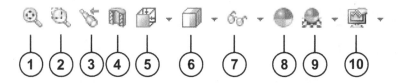

No.	Title	Description
1	*Zoom to Fit*	Adjusts the size of the model to see the object on the largest possible scale, with the presentation in full screen.
2	*Zoom to Area*	Increases the selected frame, presents it in full screen.
3	*Previous View*	Displays the previously selected view.
4	*Section View*	Shows the product cross-section of the object with one or more of the selected planes
5	*View Orientation*	Changes the orientation of the current view (front view, plan view tloris, etc.) or a number of views on the screen.
6	*Display Style*	Changes the model on screen display style.
7	*Hide/Show Items*	Hides/shows certain entities on the screen (planes, centreline, etc.)
8	*Edit Appearance*	Changes the appearance of individual model entities.
9	*Apply Scene*	Applies a certain scene to display the model.
10	*View Settings*	Enables various views (real view, shadow, perspective, etc.)

Fig. 4.5 Heads-up view toolbar

Fig. 4.6 Selection of manager commands within the
Manager window

4.2.4.1 Feature Manager

SolidWorks is, in principle, structured as a feature-based solid modeller. Each feature
that is added to the model is stored in a tree structure (Fig. 4.7). The *Feature manager*
has a built-in filter that allows us to display only the desired information.

Table 4.1 Description of the *Manager window*

No.	Title	Description
1	*FeatureManager design tree*	*Feature manager* and model structure
2	*PropertyManager*	Feature characteristics manager
3	*ConfigurationManager*	Enables working with different model configurations
4	*DimXpertManager*	Dimensions and tolerances manager
5	*DisplayManager*	Presents model visual constrains (colours, light, scenes, camera, etc.)

Fig. 4.7 Feature manager

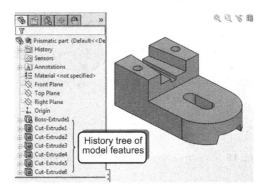

4.2.4.2 Property Manager

The *Property manager* opens up when we want to change a sketch or a feature. It contains all the current properties of the selected entity. The content of the window varies depending on the entity characteristics or the command type that is implemented (Fig. 4.8).

4.2.4.3 Configuration Manager

The *Configuration manager* is used for the production, selection and display of different model configurations that are determined in one file (Fig. 4.9).

4.2.5 Task Pane

The *Task pane* (Fig. 4.10) allows us to access the sources of the *SolidWorks* software, its libraries of often-used component parts, standard looks that can be entered into the sketch directly, etc. It works in the same way with other useful objects and information.

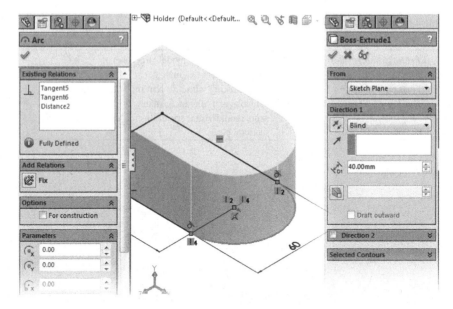

Fig. 4.8 *Property manager* in the case of arc properties and extrusion into space

Fig. 4.9 *Configuration man-ager* and different model configurations

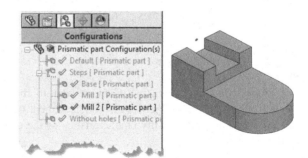

4.2.6 Status Bar

The *Status bar* displays information about what is currently active, a brief description of the commands, drawing status, etc. (Fig. 4.11).

4.2.7 Graphic Area

The *Graphic area* allows us to display and manipulate parts, assemblies and drawings. To manipulate the models we have to be aware of the possibility of inputting data from various input devices (roll bar, mouse, pen, panel, etc.) first. If we list the most

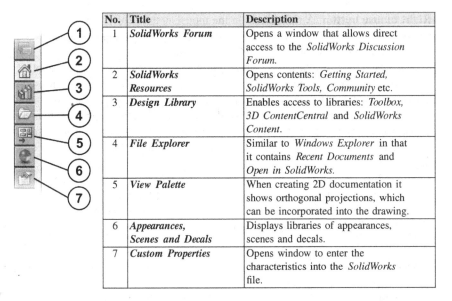

No.	Title	Description
1	*SolidWorks Forum*	Opens a window that allows direct access to the *SolidWorks Discussion Forum.*
2	*SolidWorks Resources*	Opens contents: *Getting Started, SolidWorks Tools, Community* etc.
3	*Design Library*	Enables access to libraries: *Toolbox, 3D ContentCentral* and *SolidWorks Content.*
4	*File Explorer*	Similar to *Windows Explorer* in that it contains *Recent Documents* and *Open in SolidWorks.*
5	*View Palette*	When creating 2D documentation it shows orthogonal projections, which can be incorporated into the drawing.
6	*Appearances, Scenes and Decals*	Displays libraries of appearances, scenes and decals.
7	*Custom Properties*	Opens window to enter the characteristics into the *SolidWorks* file.

Fig. 4.10 Commands scheme in the *Task pane*

No.	Title	Description
1	*Brief description*	Brief description that can be selected with the cursor.
2	*Sketch status*	Sketch status and mouse-cursor coordinates when the mouse cursor is behind the sketch.
3	*Unit System*	Information about currently selected system of units with a change option.
4	*Quick Tips*	On/off icon for quick tips.
5	*Tags text box*	Icon displays/hides the tags text box that is used for adding functions keywords.

Fig. 4.11 *Status bar* in *SolidWorks* software

commonly used actions for the mouse, the mouse's main purpose is to move the cursor on the screen, but there is also the left and the right mouse buttons and a wheel button in the middle.

- **Left mouse button**: We select individual entities with a left-click on the mouse. By pressing the *Ctrl* key button on the keyboard we can select multiple entities at the same time. Multiple entities can also be selected from the *Windows selection* window. If we drag the selected window size from left to right, we only select those elements that are inside the window. But if we drag it from right to left, all the items are selected and we can include them as a whole or just partially.

- **Right mouse button**: A right-click on the mouse is used for advanced selection and to display the properties of individual entities. By holding the right mouse button and moving the mouse we access to the pre-set shortcuts.
- **Mouse wheel button**: The mouse wheel button has multiple functions. By rotating it we can zoom in/out (see Sect. 3.3.3). By pressing it and moving the mouse at the same time we perform a rotation in space (see Sect. 3.3.2). If we also press the *Alt* button on the keyboard we can rotate around the rectangle of the screen. By pressing the *Shift* button on the keyboard and rotating the wheel button at the same time we can zoom in/out. If we add the *Ctrl* button a translation is performed (see Sect. 3.3.1).
- **Keyboard**: A keyboard is an important addition to the mouse, where we can easily adjust a pre-set shortcut for every key. The following keys and their combinations are commonly used in *Solidworks* software in addition to the classic combinations that are specific to the *Windows* environment:

  - *Esc* Command cancel, unselect all
  - *Spacebar* View orientation menu
  - *Ctrl + Spacebar* View selector
  - *Z* Zoom out
  - *Shift + Z* Zoom in
  - *F* Zoom to fit
  - *S* Shortcut bar
  - *Ctrl + B* Rebuild the model
  - *Ctrl + R* Redraw the model
  - *Ctrl + Z* Undo

4.2.8 SolidWorks Options

In addition to the presentation of the user interface the main settings of the *SolidWorks* software are presented below. To access the optional window we select *Tools > Options...* or we click on the icon in the menu bar. Two tabs are available:

- *System Options*, valid for the entire software,
- *Document Properties*, where the properties for each document are set separately (part, assembly, drawing).

4.2.8.1 System Options

System options are recorded in the register of the software (the database of the basic data) and are not part of the document, while changing them affects all documents (current and future). Figure 4.12 shows a dialog box to set the system options.

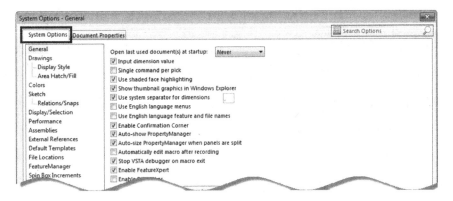

Fig. 4.12   *System options* dialog box

The following options are available:

- *General* Specifies the general system options such as enabling the performance feedback option and the *ConfirmationCorner*.
- *Drawings* Sets options for all drawings.
- *Colours* Sets the colours in the user interface: backgrounds, drawing paper, sketch status, dimensions, annotations, etc.
- *Sketch* Sets the default system options for sketching.
- *Display/Selection* Specifies the options for the display and selection of edges, planes, etc.
- *Performance* Changes to these settings do not affect the documents that are already open.
- *Assemblies* Sets assembly options, including options for *Large Assembly Mode*.
- *External References* Specifies how part, assembly and drawing files with external references are opened and managed.
- *Default Templates* Specifies the folder and template file for automatically created parts, assemblies and drawings.
- *File Locations* Specifies the folders to search for different types of document. The folders are searched in the order in which they are listed.
- *FeatureManager* Sets the *FeatureManager* design tree options.
- *Spin Box Increments* Sets the increments for *SolidWorks* spin boxes.
- *View* View displayed on the individual sketch.
- *Backup/Recover* Sets frequency and folders for auto-recovery, backup, and save notification. The auto-recovery and save notification are controlled by a specified number of minutes.
- *Search* Sets options for *SolidWorks File* and *Model Search*.
- *Collaboration* Specifies options for a multi-user environment.
- *Messages/Errors/Warnings* Controls whether certain warning messages are displayed. You can restore messages that have been suppressed.

Fig. 4.13 *Document properties* dialog box

4.2.8.2   Document Properties

Document Properties are set separately for each document and are recorded in the file. They (Fig. 4.13) are activated by selecting the *Document Properties* tab. Some features are available for all types of documents (part assembly and drawing), while others are tied to a specific document type.
We can manage the following properties:

- *Drafting Standard* Specifies the document-level overall detailing drafting standard, and rename, copy, delete, export, or load saved custom drafting standards. Available for all document types.
- *Annotations* Specifies document-level drafting settings for all annotations. Available for all document types.
- *Dimensions* Specifies document-level drafting settings for all dimensions. Available for all document types.
- *Centerlines/Center Marks* Sets drawing document properties for centerlines or centre marks. Available for drawings only.
- *Drawings* Sets options for chamfers, slots, and fillets for use with the *DimXpert tool*. Available for drawings only.
- *Virtual Sharp Display* Sets display options for virtual sharps. Available for all document types.
- *Tables* Specifies document-level drafting settings for all tables. Available for all document types. Some settings are only available for drawings.
- *View Labels* Specifies document-level drafting settings for all view labels. Available for drawings only.
- *Detailing* Specifies document-level drafting settings for detailing options. Available for all document types.
- *Grid/Snap* Displays a sketch grid in an active sketch or drawing and set options for the grid display and snap functionality. Available for all document types.

- **Units** Specifies document-level properties of units. Available for all document types.
- **Line Font** Sets the style and weight of lines for various kinds of edges. Available for drawings only.
- **Line Style** Creates custom line styles and applies them to edges. Available for drawings only.
- **Line Thickness** Sets the line weights in an active document that work best with your printer or plotter. These settings are saved for all open *SolidWorks* documents. Available for drawings only.
- **Model Display** Changes colour options for model display. Available for parts and assemblies.
- **Material Properties** Sets crosshatch options and material density for the active part. Available for parts only.
- **Image Quality** Specifies quality options for image display. Available for all document types.
- **Sheet Metal** Specifies sheet-metal options. Available for all document types. Options vary depending on whether you are working with a part, assembly or drawing.
- **Plane Display** Specifies colour, transparency, and intersection options for plane display. Available for parts and assemblies.
- **DimXpert** These options define whether DimXpert uses *Block Tolerances* or *General Tolerances* on dimensions that do not contain tolerances. Available for parts only.

4.3 Siemens NX PLM

NX Siemens PLM is an advanced software package that in addition to 3D modelling comprises several dedicated modules that facilitate the work of the various stages of the very complex formation of products. It provides advanced solutions of conceptual and 3D design and it manufactures the documentation. Additional modules are, for example, for structures and fluid simulations and modules for the manufacturing technology of products. It is one of the most expensive and complicated software packages. It works on the following operating systems: *Windows*, *Linux* and *Mac OS X*.

4.3.1 Introductory Window

At start-up the *NX* modeller for the first interface's window opens (Fig. 4.14). As usual we can choose between templates for a new document or we can open an existing one. When selecting a new document a new window opens up, where we

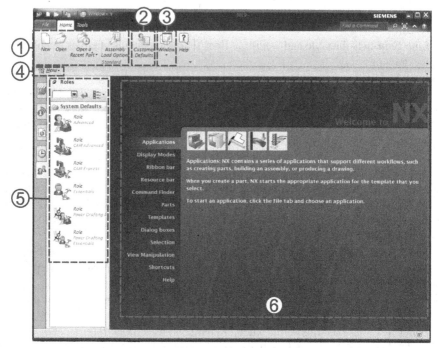

No.	Title	Description
1	*Document tabs*	Initial set of commands for documents (*New, Open, Open Recent* , etc.).
2	*Customer Defaults*	User Settings (standards).
3	*Window*	Choice of a document among other open documents.
4	*Menu*	The main menu with all the commands and settings.
5	*Roles*	User profiles.
6	*Welcome window*	Basic information clearly shown (very useful for beginners).

Fig. 4.14 Introductory window

can select a document type. One of the main features of *NX* is that it has a single format for all the main types of document (part, assembly, drawing) which is *.prt.

After selecting the desired document (in our case part-model) we move on to the user-interface module (*Modelling*) (Fig. 4.15). Version *NX9.0* is the first version to use an interface type *Ribbon*. Figure 4.16 shows the basic design of the interface.

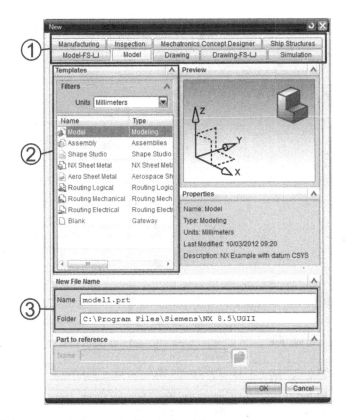

Fig. 4.15   Creating a new document and a set of modules

4.3.2 Manipulating the View in the Graphic Window

The easiest way to manipulate the view of the 3D object in the graphic window is using the 3D mouse (*3D Device*), which has six axes in combination with the regular three-button mouse. If only the regular mouse is available we can accomplish this using combinations of the mouse buttons as shown in Fig. 4.17 (MB1 = left button, MB2 = right button and MB3 = middle button).

4.3.3 Ribbon Bar

The interface menus in the *Ribbon* type were first introduced in *NX* version 9.0 (Fig. 4.18). It is a pre-set, user-friendly structure of commands, where we can achieve what we want with the least number of clicks. The commands are distributed in individual sections or strips by their similarities on the top of the user interface.

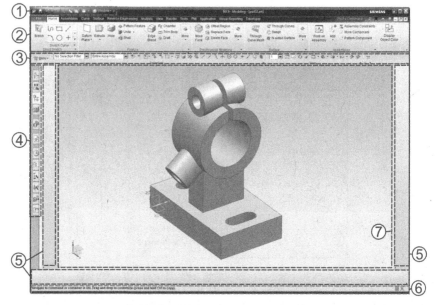

Fig. 4.16 Basic graphical interface of the *NX* modeller

No.	Title	Description
1	*Quick access tool bar*	Contains commonly used commands such as *Save* and *Undo* .
2	*Ribbon bar*	Organizes commands in each application into tabs and groups.
3	*Top border bar*	Contains the *Menu, Selection Group, View Group*, and *Utility Group* commands.
4	*Resource bar*	Contains navigators and palettes, including the *Part Navigator* and the *Roles tab*.
5	*Left, right and bottom bars*	Displays the commands you add.
6	*Cue/status line*	Prompts you for the next action, and displays messages.
7	*Graphic window*	Lets you model, visualize, and analyse models.

The *Tab* component combines different groups of commands and tools (modelling commands, assembly manipulating, analysis, etc.). These sets allow us to see individual **Groups**, each containing related commands (commands for sketching, various extrusions, synchronous modelling, etc.). In the **Toolbar** *options* settings we can turn on/off any of these orders. The command line **Command finder** is very useful (Fig. 4.29). Here we can find any command when inserting a rough approximation of the command name or the function we want to perform. In **Full screen** view tool bars and menus are temporally hidden (Fig. 4.19).

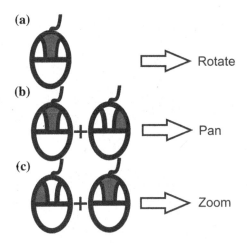

Fig. 4.17 Using combinations of the three mouse buttons and moving the mouse at the same time it is possible to manipulate the view of the space with six degrees of freedom: Button the mouse and commands: middle—rotate (**a**), middle and right—pan (**b**), middle and left—zoom (**c**)

No.	Title	Description
1	*Tab*	Organizes commands into groups of related functions in each application.
2	*Group*	Organizes commands by function in each tab.
3	*Toolbar options*	Enables the turn on/off commands in each group.
4	*Command finder*	Finds commands.
5	*Full screen*	Maximizes the screen space.
6	*Minimize ribbon*	Collapses the groups of ribbon tabs.
7	*Help*	Displays on-context *Help*

Fig. 4.18 *Ribbon bar*, first introduced in the *NX 9.0 version*

4.3.4 Top Border Bar

At the left end of the *Top border bar* there is a window to the main *Menu*, where all the settings and commands of the NX software are listed in the drop-down mode (Figs. 4.20 and 4.21). The specific geometry filters are listed from left to right, which enables us to select a specific geometry with the cursor. We can choose between curves, shapes and objects, etc. Next, follow commands that are connected to the perception of the geometric properties and *Shaping*, for example, the centreline of

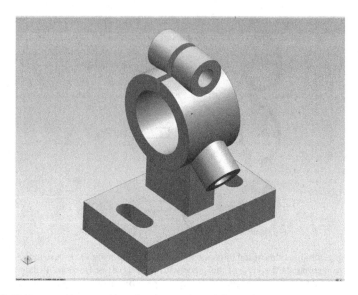

Fig. 4.19 Full-screen view provides a better overview of the model. Menus are hidden

No.	Title	Description
1	*Menu*	All the existing commands and settings are located here.
2	*Selection group*	Using filters we can set the selection of specific geometric entities (line, shape, object, etc.).
3	*Snap options*	Various commands for perceiving geometric points (centreline of circle, end points).
4	*Layer options*	Selection of different layers.
5	*View options*	Model view manipulation in *Graphic Window*.

Fig. 4.20 Components of the *Top border bar*

circles, the endpoints of a line, the centre of a line, etc. This function can be switched on/off.

4.3.5 Resource Bar

The vertically aligned *Resource bar* contains important sets of structural trees, like *Part Navigator*, which features a sequential presentation in modelling the

Fig. 4.21   All the settings and commands of the *NX* software are gathered in the main menu

parts, *Assembly Navigator*, *Assembly* ***Constraints Navigator*** and various libraries (Fig. 4.22). It should be noted that *NX* also provides a ***History-Free Mode***. In this mode the historical traces of steps, links and references between features are obscured. In the case of the history-free mode we use the principle of ***Synchronous modelling***. This toolbar also contains an integrated internet browser, personal profiles and some other useful tools.

Switching between the first and second modes is via a mouse right-click on the empty space inside the ***Part Navigator*** window (Fig. 4.23). When the menu is opened we uncheck the box in front of the ***Timestamp order*** line to change the display mode. If we want to switch between the historical and non-historical modes, we position the cursor on the title ***History mode*** on the top of the ***Part Navigator*** tree structure and right-click the mouse button. We need to be aware that switching from the historical to the non-historical mode or the reverse means losing the entire history of the modeled part.

4.3.6 Radial Tool Bar/Shortcut

The *Radial tool bar* enables the rapid selection of the commands. It is activated by pressing the ***Shift*** and ***Ctrl*** keys on the keyboard and one of the three mouse buttons at the same time (***MB1, MB2*** and ***MB3***) (Fig. 4.24). Based on the default settings,

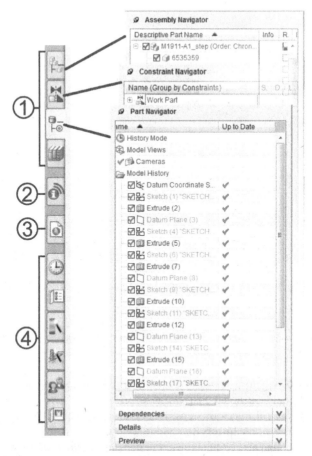

No.	Title	Description
1	*Navigators*	Set of features trees and other components in the parts, assemblies and libraries.
2	*HD3D tools*	Tools for the display and the interaction with data directly on the 3D model.
3	*Integrated Web browser*	Interface for internet access integrated into the *NX* software.
4	*Palettes*	Access to templates set by user, system visualization and materials.

Fig. 4.22 The vertical alignment of the *Resource bar*, where various menus such as *Assembly Navigator, Part Navigator*, profiles, etc can be found

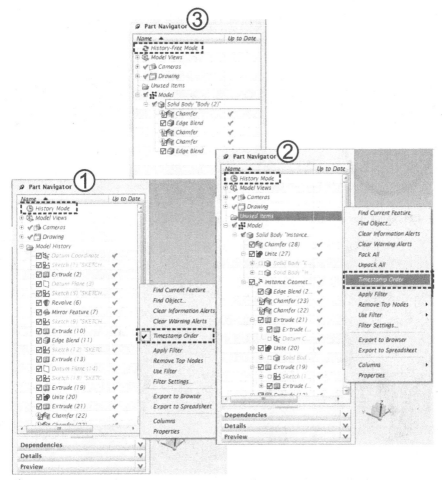

No.	Description
1	Features are arranged by date in the order of their generation.
2	The structure labels the hierarchical placement of features and elements based on their mutual reliance. This is still in historical mode.
3	Non-historical mode. Steps are not traced. Only the 3D objects are left.

Fig. 4.23 Three different *Modes* of the *Part Navigator* when modelling the same 3D part

we can use the third mouse button to display a fourth quick menu. In this menu, we can choose among different views of the parts, such as: a wire frame, a boundary representation with shadowing, renders, etc.

We can display the last quick menu using a mouse right-click (*MB2*) on the desired position of the part. Only commands that are related to the selected geometric entity or feature are available. This is graphically shown in Fig. 4.25.

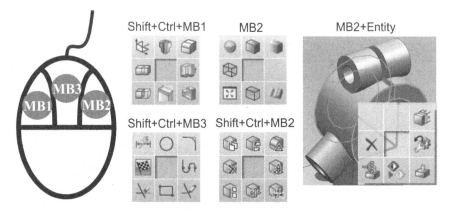

Fig. 4.24 Quick way of activating the radial toolbars

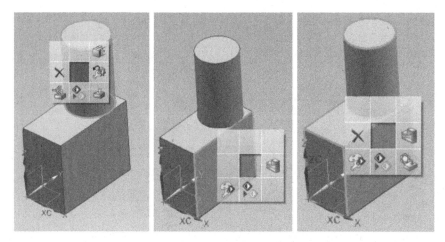

Fig. 4.25 Demonstration of different commands available using the selection of various geometrical entities using the right mouse button

4.3.7 Keyboard Shortcuts

We also have keyboard shortcuts available. Some of the pre-set shortcuts are shown in Table 4.2. A full list of pre-set shortcuts is available in the software: *Menu > Information > Custom menu bar > Shortcut key*.

Table 4.2 Pre-set keyboard shortcuts

No.	Key on the keyboard	Task
1	*Home*	Orients geometry in trimetric view
2	*End*	Orients geometry in isometric view
3	*Ctrl+F*	Fits geometry to the graphic window
4	*Alt+Enter*	Switches between standard and full screen view
5	*F1*	Displays *Help* on context
6	*Ctrl+Shift+S*	Displays information window

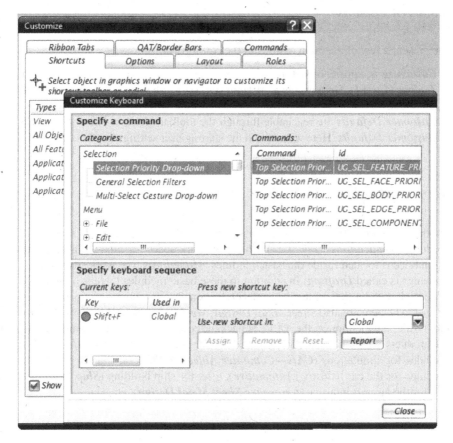

Fig. 4.26 Menu for customizing the keyboard shortcuts

The user can customise the shortcuts in the *Customize Shortcuts* menu: *Menu > Tools > Customize > Shortcuts > Keyboards* (Fig. 4.26).

Fig. 4.27 The *Application tab* consists of various user modules

4.3.8 NX Options

We present a few of the main options of the *NX* software.

- *Customize* is a group of settings that can be displayed through the *Quick Access Toolbar* (*Ctrl+1*). Here we have different settings for the window, commands and shortcuts.
- *Customer Defaults* are reachable through the toolbar *Menu > File > Utilities > Customer Defaults*. Here we can find the starting user settings for the commands.

4.3.9 Application Tab

On the *Ribbon bar* there is a menu for selecting different special work spaces and modules (*Applications*) (Fig. 4.27). The main work space for modelling parts and assemblies is located inside the basic modules (*Modelling*). The work space for 2D drawings is called *Drafting*. In addition to these basic modules there is a full range of special modules that are available, depending on the licensing restrictions. As part of the CAD there are also *Sheet Metal Design*, *Mechanical Routing*, *Shape Studio,* etc. As already mentioned, the *NX PLM* package contains tools for the completion of the products We should also mention modules for manufacturing (*Manufacturing*), modules for simulations (*CAE—Computer Aided Engineering*), a set of specific modules for the car industry (*Automotive*), a set for ship building (*Ship building*), one for the aircraft industry (*Aerospace Sheet Metal Design*), etc.

4.3.10 Synchronous Modelling

The *Siemens NX PLM* modeller also has a set of features for so-called *Synchronous modelling* (Fig. 4.28). This is an intelligent method to manipulate the geometry, which does not require the history of the object modelling because the features automatically identify geometric parameters and give the change options. *Synchronous modelling* is a set of useful commands, which are part of advanced modelling.

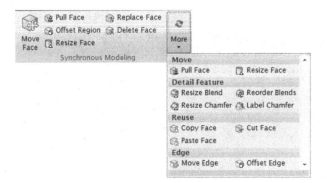

Fig. 4.28   Commands set for synchronous modelling that are part of the *Ribbon bar*

4.3.11 Command Finder

We would particularly like to emphasise the ***Command Finder*** (Fig. 4.18, number 4).
It is the most useful part of the interface for new users. We type the word that is related
to the desired feature into the search window and the software itself displays a set
of commands that can be used. The user then selects the desired commands, which
automatically opens in the menu and is then activated (Fig. 4.29).

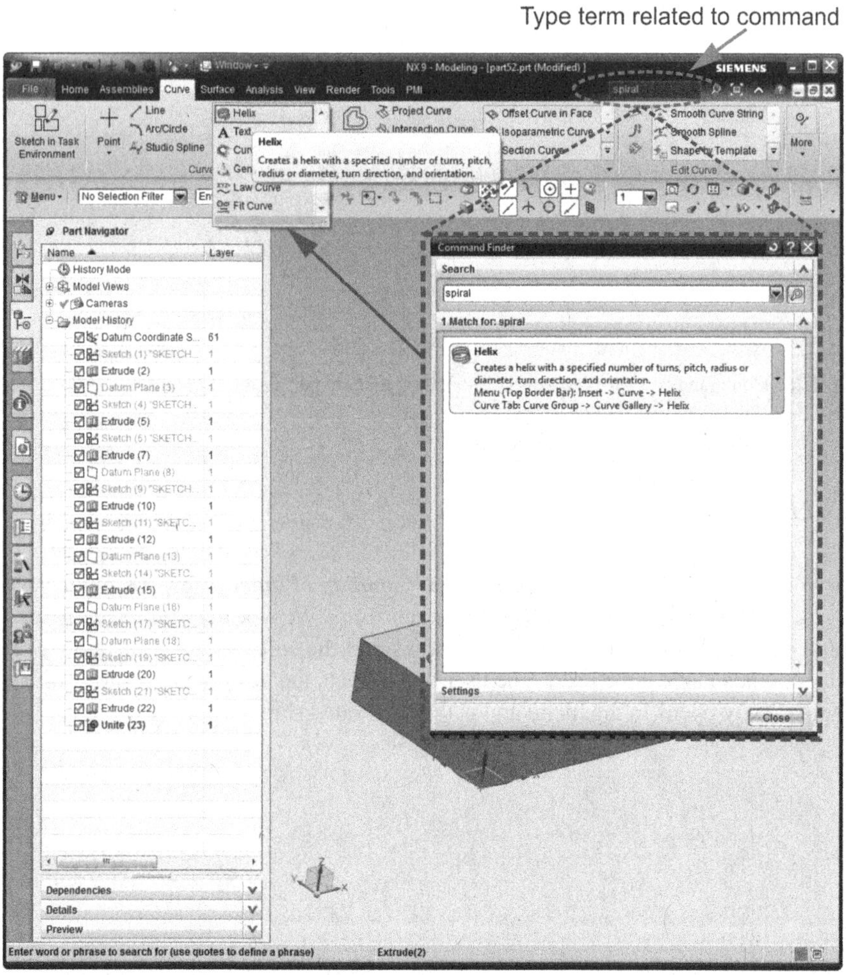

Type term related to command

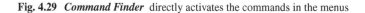

Fig. 4.29 *Command Finder* directly activates the commands in the menus

Chapter 5
Extrusion

Abstract Extrusion is the basis for forming prismatic bodies. The first sub-chapter deals with extrusion in the case of industrial examples in order to make the reader familiar with why this feature appears as the basic modelling feature. The concept of defining a base sketch on a plane and forming a definite (or indefinite) vector of extrusion is described. The chapter is concluded with a couple of typical extrusion examples.

There are many examples of extrusion, for example, rolling, the production of profiles or the steel beams, frames and trusses used in construction. Extrusions are also often used to form plates, wheels, profiles and various plane blocks. We make sure that the profile is along the length of the extrusion and as a result we significantly simplify the manufacture of various types of pro-ducts, such as profiles, extruded polymers and aluminium profiles. We also use extrusions with rolling or linear milling profiles, which are often used in steel or wooden trusses. In order to understand the suitability of this feature, we will present some typical applications of extrusion below.

5.1 Manufacturing Technology

The best-known type of extrusion is the rolling of different profiles. Old shapes of profiles were restricted to a simple method of rolling, like that shown in Fig. 5.1. Hot-rolling profiles used to be made with a slope flange or parallel flanges with rolling, while the latest ones have all the sides in parallel, as shown in Fig. 5.2.

All the examples in Figs. 5.1 and 5.2 are typical for rolling profiles. Extrusion has recentl y been used at crystallization (polymerization) temperatures. This is very common for polymers, aluminium, magnesium, bronze, copper, etc. These materials are plastically deformed between 100 and 750 °C. Figure 5.3 shows a variety of profiles that are used for different purposes.

For the simulation of drilling it is necessary to use revolving. In this case we make a basic sketch with the full length of the drilling hole and complete the classic drill with a slope at the end of 120°. Revolving is used for a blind hole. If we have a

J. Duhovnik et al., *Space Modeling with SolidWorks and NX*,
DOI: 10.1007/978-3-319-03862-9_5, © Springer International Publishing Switzerland 2015

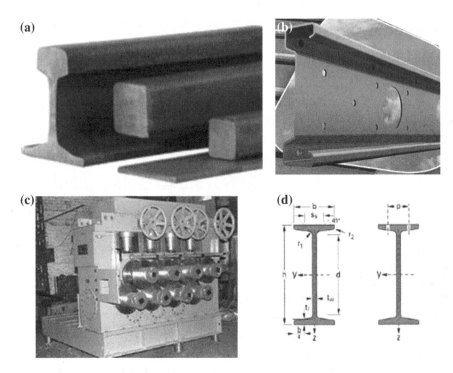

Fig. 5.1 Cross-sections of different hot-rolling profiles and rolling principles: profiles (rail, square, flat...) (**a**), special profiles manufactured with bending sheets metal (**b**), machine for continuously bending (**c**), standard I-profiles (hot rolling) (**d**)

Fig. 5.2 Rolling principles process (**a**) and hot-rolling profiles with slope flanges (**b**)

through hole or a blind milling hole with a flat bottom we need to model this shape with a typical circle, from which we form an object that is produced from the different material and that represents a round hole. A through hole is presented in Fig. 5.4c. The blind hole is shown in Fig. 5.4a.

Fig. 5.3 A variety of profiles for different purposes and matrices by extrusion

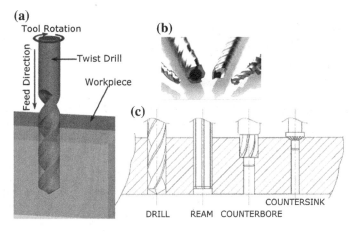

Fig. 5.4 Full and part-length hole and the drilling principles. Blind bore (**a**), through bore and special bore—tool in (**b**), drawing in (**c**)

We also obtain extruded shapes when milling lengths of wood, for example, parquet (classic, profiled boards), picture frames, window or door frames and other wooden claddings). Such shapes are obtained using a mill with a profile and continuous motion. The shapes and technological processes are shown in Fig. 5.5.

We often utilize extrusion technology when forming prismatic objects, which can be finished with flat faces (cut-offs or cut-outs) or with a free-form shape.

5.2 Modelling Prismatic Objects

The easiest way to model prismatic objects is by determining the basic plane for sketching, for which we need to determine the direction and length of its movement with a vector feed in a specific way. The length of the movement defines the height

Fig. 5.5 Wooden profiles produced by milling

Fig. 5.6 Placement of the model in space and the orthogonal projections on the basic plane for a 2D demonstration of a product or an object

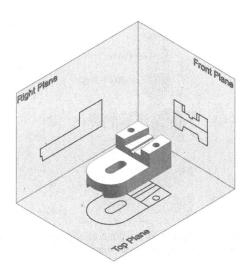

of the prismatic object, which has a pre-determined face. Movement in the direction of the third axis is called extrusion. In this way a 3D model is formed.

5.2.1 Object Formation in Space

For the formation of an object in space, we first need to determine in which space and in which way we would like to place the modelled object. To obtain a clear orientation it is best if we place the object in space in the same way that it appears in nature (Fig. 5.6). If the product normally stands vertically, the model should be vertical too;

Fig. 5.7 Basic planes and coordinate systems of a rectangular coordinate system

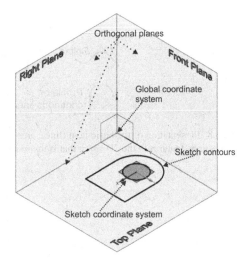

however, if it is usually in a horizontal pose, we model it horizontally. Products that are usually placed obliquely to the horizontal and vertical directions are modelled in this way only in specific cases. Other special cases include objects presented in space in such a way that they can be seen on machines or on devices. The proper choice of the initial placement of an object or product in a space significantly contributes to an easier understanding of the model and the production of the technical documentation later on. With a proper virtual placement of the product model we contribute to an understanding of the shape and also the function itself.

5.2.2 Basic Sketches on the Basic Plane

A basic sketch is intended to present the basic plane of a model. The sketching plane is usually in 2D and it provides the argumentation of the initial planes and indirectly of the space we will use to present the model or the object. In the sketching plane we make the basic sketch in 2D, which we subsequently stretch by using the appropriate tools in the direction of the vector parallel edges in the direction of the third coordinate, which we have not used before. This procedure is called extrusion and gives us a model in 3D. The basic planes of the sketch are the front view (x–y), the top view (x–z) and the side view (y–z). We can also make optional planes in space (Fig. 5.7).

We can choose the basic plane by ourselves, but it should provide the best demonstration of the model. To properly arrange the professional work and to ensure that there is comprehensive information relevant for the prismatic parts we normally use the (x–y) plane, which we call the front view.

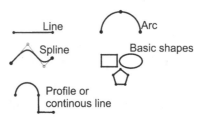

Fig. 5.8 Presentation of the basic item (lines, arcs, splines, basic shapes and their combinations) that occur as features in the 2D space that is the space for making the sketch

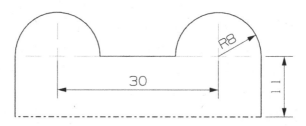

Fig. 5.9 Example of a basic sketch for a model that is implemented with extrusion from the (x–y) plane

We produce the basic sketch in the selected sketching plane (front view) that we are going to stretch in space using the extrusion function. It is important to determine the size and the position of each of the sketch elements (line, arc, curve, etc) (Fig. 5.8). This can be made by using dimensions and relations to obtain a completely defined sketch. We must always determine the sketch position in the selected plane. The easiest way to accomplish this is to position the characteristic element of the sketch in the local coordinate system and select one of the axes as an initial axis (Fig. 5.9). In the case of block-square shapes this is either the left or the right edge. In the event of an axisymmetric model, one of the axes is equal to the position and the direction of the centre line of the model (Fig. 5.10).

In the case of symmetric models we can produce a comprehensive sketch by drawing a part of the geometry first, which can then be mirrored through the corresponding axis (*Mirror Feature*).

Among the individual sketch elements (point, straight line, arc, circle, etc.) various relations can be added between them. These relations are marked as horizontal, vertical, tangential, coincidence, parallel, and perpendicular (Fig. 5.11). The basic sketch must always be dimensioned so that the dimensions of each element are clearly defined. The process for the sketch is: (1) we make a rough sketch considering the various relations, and (2) we introduce a dimensional size for features at the same time as a function of the sketch dimensioning (*Dimension*).

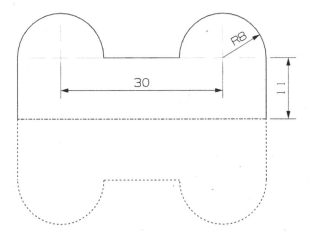

Fig. 5.10 Example of a basic sketch of a symmetric model that is implemented by mirroring through a pre-set edge that represents the mirroring axis

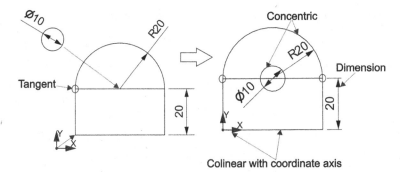

Fig. 5.11 Examples of the relations or constraints between individual features of the sketch element

Some modelers are able to perform the extrusion of free-form curves or even of a free-form shape (Fig. 5.12). This function is very useful for products and models with a free-form shape.

5.2.3 Extrusion of a Basic Sketch on a Plane Into Space

The basic sketch is composed of a contour and a basic line (centre line, auxiliary line, generating line, etc.). The auxiliary line clarifies the contour and does not expand into space. A high-quality sketch provides a closed and complete inventory of the 2D modelling features and needs to present a clear shape for the auxiliary planes to the engineer. Only a closed contour enables the formation of a full solid model using extrusion. In contrast, shape models can present both open and incomplete contours.

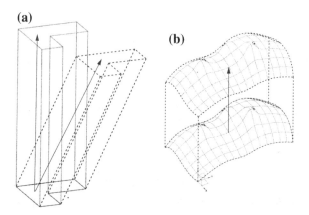

Fig. 5.12 Extrusion of a closed contour or basic sketch into the direction of a third axis using a vector of optional direction enables a solid model: extrusion of planar sketch (**a**) and extrusion of free form shape (**b**)

Fig. 5.13 The extrusion of a sketch into a space with an incomplete contour is possible by using a shell function that automatically creates the basic plane in a plane

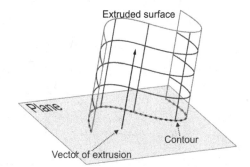

If we use open and incomplete contours to form a shell, we obtain a thin-walled model. Such a thin-walled model on a basic plane already represents a plane that is bounded with a contour. This example is used with sheet products or products made of polymer materials.

Extrusion can be defined: (1) the direction of an extrusion vector and (2) the extrusion length (model height). The better modelers have different options for defining the length of the extrusion. The most common option is (*StartValue*), (*Direction1*) and (*End Value*), (*Direction2*) that are determined by distances relating to the plane the sketch is in. Another option for extrusion relates to the existing shape/surface (*UntilNext*), (*Until Selected*). Using these principles we limit and thereby adjust the extrusion length of a certain entity, for example, a free-form shape (Fig. 5.13). Using a *Through All* function the extrusion is performed in a selected direction through the whole object (Fig. 5.14).

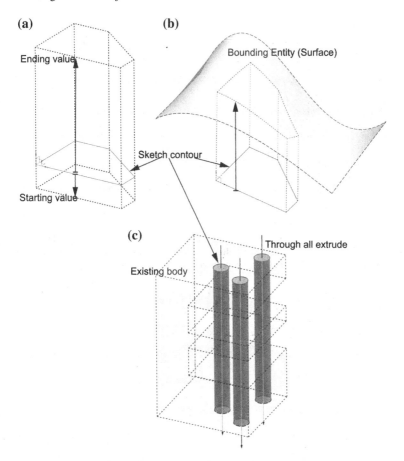

Fig. 5.14 Examples of length-determination options in extrusions: vector ending value (**a**) bounding entity surface (**b**) and through all extrude (**c**)

5.2.4 The Formation of Complex Shapes and Details Using Extrusions

Extrusions can be combined into a more complex shape by Bool's algebra. The modelling process for a complex form and the related details has to be defined at the beginning of the 3D modelling of the object. The modelling of a product needs to be planned in a sense that the working steps are set reasonably and that we achieve the complete shape with the minimum number of steps. The most important factor is the formation and the positioning of the basic sketch. From its position and shape the sequence of subsequent steps for the formation and assembly of different features into the final shape of a model or a product is determined. In this process we must logically implement the composition of the prismatic form and at every step check whether it is relevant to the aggregation or removal of an individual prismatic form

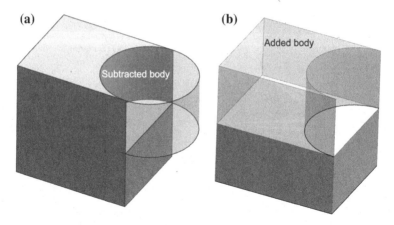

Fig. 5.15 Two modelling approaches. (**a**) Two prismatic bodies—removing a cylinder from a box and (**b**) the aggregation of two prismatic bodies—the addition of a box with a prismatic body that has a basic sketch with a clipped circle on one edge

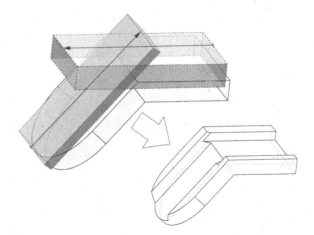

Fig. 5.16 Two prismatic bodies that are combined after detailed space positioning in such a way that one is rotated around the axis that is determined as the edge of a composed model

or even dispose of parts of the basic sketch. We can see two different approaches that lead to the same result in Fig. 5.15. The difference between the two approaches is in the number and the complexity of the steps (Fig. 5.16). The use of an approach that is too complex can result in complications in the subsequent modelling process.

The individual steps of modelling a prismatic model from Fig. 5.17 are below. The modelling process can be summarized in the following steps—features (Fig. 5.18):

1. Basic form
2. Cut-out from the front side on the upper level
3. Cone cut-out on the back side on the upper level

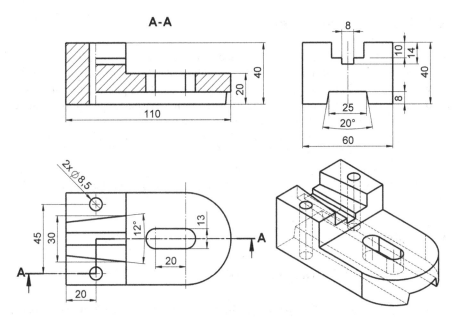

Fig. 5.17 Example of a model of a handhold $110 \times 60 \times 40$ mm

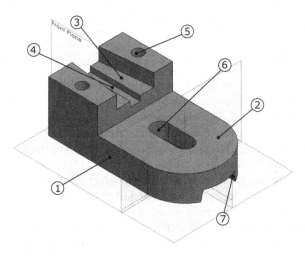

Fig. 5.18 Features on the handhold model $110 \times 60 \times 40$ mm

4. Groove of 8×4 mm
5. Through hole $\phi8.5$ mm
6. Slot $\phi13 \times 20$ mm
7. Slot 25×8 mm on the bottom level
8. Final model

5.3 Modelling in SolidWorks

The handhold model from Fig. 5.17 is modelled using features from Fig. 5.18. It is useful to consider the manufacturing technology that is going to be used later on when manufacturing the product during the 3D modelling phase. The handhold model is made only with cutting technology on a milling machine, which is presented in Sect. 5.1. The base (*Boss/Base*) needs to be made first, from which we deduct the volume by using Bool's operation countdown and we continue to gradually build the final model.

We create a new part by selecting *File > New*. When a new dialogue box opens, from which we select the option *Part* and we obtain a working space for the modelling of individual parts, see Fig. 5.19.

5.3.1 Basic Form

5.3.1.1 Placement of an Object in Space

The placement of an object in space is made by the placement of the basic sketch. We create the basic sketch on one of the main projection planes. In our case we selected the top-view plane (Fig. 5.20).

5.3.1.2 Basic Sketch

We draw the basic sketch (Fig. 5.21) on the selected top-view plane. We start drawing the lines and arcs that form the contour of the cross-section or shape that we would like to extrude in the space. The *Mates* function is used to define the connections of individual entities. The size of the individual entities is determined by the sketch dimensioning (*Dimension*). It is essential to precisely determine the drawing position on the selected plane. The simplest way is to position one characteristic element of the drawing into the local coordinate system and select the *x* or *y* axis as the initial axis. In the case of square shapes this would be the left or right edge. When modelling symmetric shapes the initial shape is placed in such a way that the centre line lies on one of the main projection planes. To have a precise inventory of the model's shape and its position in space it is necessary for the sketch to be *fully defined*.

5.3.1.3 Extrusion Into a Space

We build a 3D object with the extrusion of the planar section curve in a sketch (*Insert > Boss/Base > Extrude*). During the modelling of a volumes model it is important that the contour is closed and for the sketch to be completely defined.

To extrude into a space we select suitable parameters and set their values (Fig. 5.22). In our case we normally perform the extrusion in a sketch plane using a height of 40 mm.

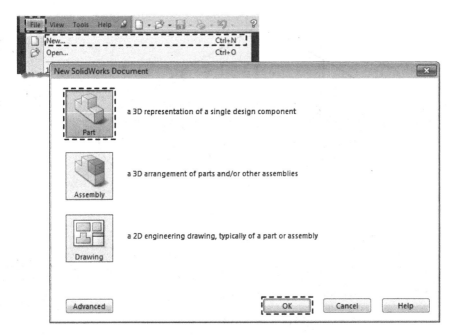

Fig. 5.19 Dialogue box for the selection of a new document type

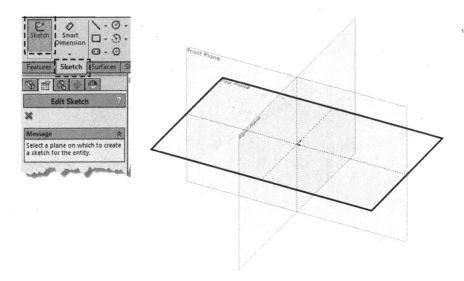

Fig. 5.20 Selecting a sketch plane to draw the basic sketch. The top-view plane is selected

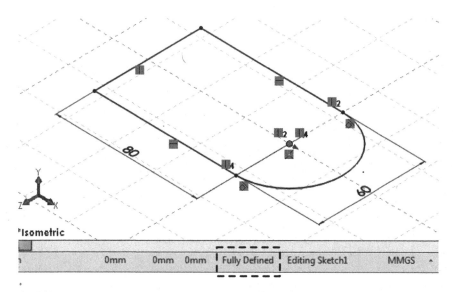

Fig. 5.21 Fully defined basic sketch in a top-view plane

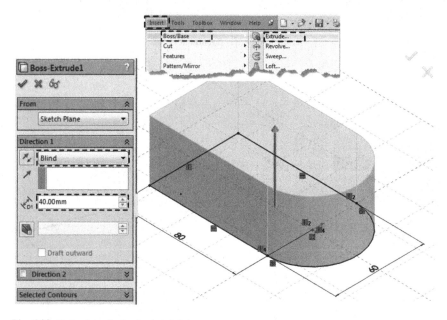

Fig. 5.22 Extrusion of a basic sketch into a space

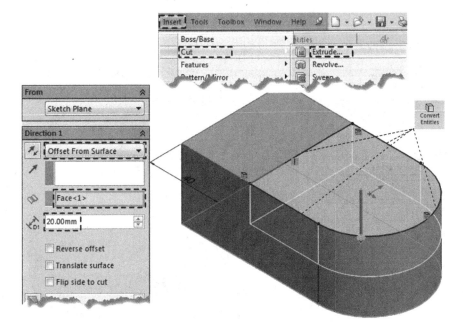

Fig. 5.23 Creation of a cut-out on the front top part

5.3.2 Cut-Out on the Front Top Part

A cut-out on the front top part is performed with an extrusion so that from the basic volume we produce a prismatic form. The existing contours are converted into the sketch elements with the command *Convert Entities*.

The cut-out is performed with the command *Insert > Cut > Extrude*. Figure 5.23 shows the production of a cut-out on the top part. We implement the extrusion where the closed plane is 20 mm from the initial (bottom) plane.

5.3.3 Cone Cut-Out 30 × 10 mm on the Back Top Part

We create a sketch of the cone cut-out on the top surface. As part of this process we adopt two existing edges (*Convert Entities*). However, because we have a symmetrical sketch in this case we can use the relation *Symmetric*.

The extrusion into a space is performed with the command *Cut*. We normally start the process in the sketch plane with a blind extrude on a sketch plane with the value of 10 mm (Fig. 5.24).

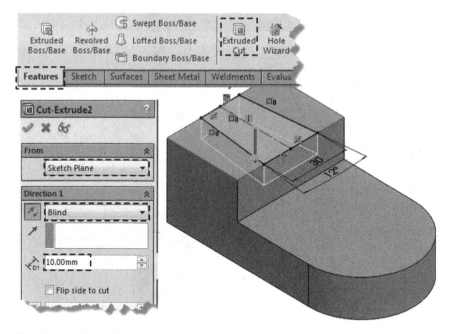

Fig. 5.24 Modelling of cone extrude on the back top part

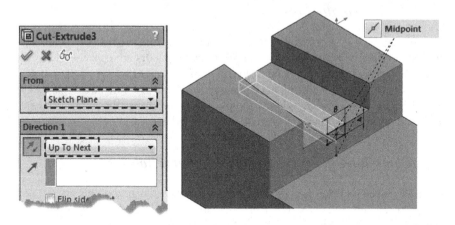

Fig. 5.25 Groove modelling 8 × 4 mm

5.3.4 Groove Modelling 8 × 4 mm

A groove of 8 × 4 mm can be produced in different ways. Figure 5.25 shows the process where the sketch (a rectangle of 8 × 4 mm) is drawn on a side plane. We do

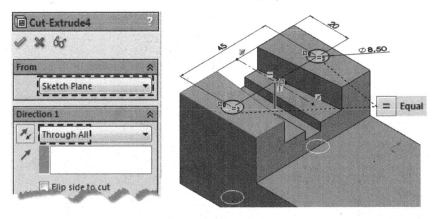

Fig. 5.26 Modeling through bore ϕ8.5 mm

this using a centre line, which extends from the centre (*midpoint*) of the top edge to the centre of the bottom edge. With the extrusion into a space we use the option *Up To Next*.

5.3.5 Creation of Through Holes ϕ8.5 mm

We only use the extrusion into a space in the case of through holes. A blind hole is modelled by using the revolving feature (see Chap. 6) or a specific function of the manufacturing bores, because of its specific shape in the bottom of the bore (*trail cutting edge bore* on the cutting plane).

Figure 5.26 shows the process of producing through bores with a diameter of 8.5 mm. We draw two circumferences, which are symmetrical (*Symmetric*) regarding the centre line that goes through *Midpoints* of the side edges, on the top surface. We dimension (*Dimension*) one of the circumferences and connect the other with the relation *Equal*.

When extruding into a space we select the option *Through All*.

5.3.6 Slot Modelling ϕ13 × 20 mm

We can produce a groove in a similar way to a hole ϕ13 × 20 mm (Fig. 5.27). The SolidWorks software contains a tool for sketching grooves (*Sketch > Straight Slot*), which is aligned with a centre line rounding on the front part (*Coincident*). The size is defined by dimensioning (*Smart Dimension*).

The extrusion into a space goes through the entire volume (*Through All*).

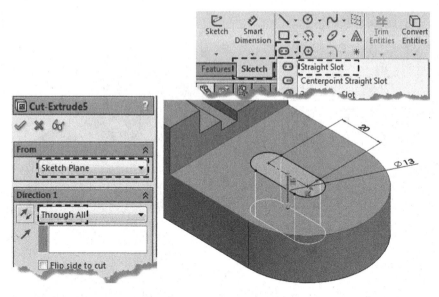

Fig. 5.27   Slot modelling $\phi 13 \times 20$ mm

5.3.7 Grove Modelling 25 × 8 mm on the Bottom Part

We model the groove of $\phi 13 \times 20$ mm on the bottom part by drawing the groove shape on the bottom surface. At the same time we adopt the front and the back edges (***Convert Entities***). We bound the width of the groove using two lines that are symmetrical to the centre line and define them with a dimension of 25 mm.

Since the side groove is angled at 10° we use the function ***Draft On/Off*** to extrude into space and set the angle to 10^{circ}. If we wish to angle the model only on the side faces we need to shift the start of the extrusion (***From***) parallel (***Offset***) by 8 mm into the model. We set the extrusion into space using the option ***Through All*** (Fig. 5.28).

5.3.8 Final Model

Figure 5.30 shows the final model of the handhold, its feature structure (***Feature Manager Design Tree***) and mass characteristics (***Mass Properties***). We set the material properties using a right mouse click on the feature ***Material*** and a left mouse click on the option ***Edit Material***. From the window that opens we select the desired material (Fig. 5.29). The model characteristics (volume, mass, surface, etc) can be seen by activating the group of functions ***Evaluate*** and selecting ***Mass Properties***.

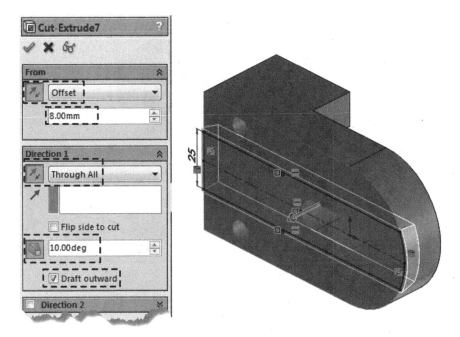

Fig. 5.28 Grove modelling 25 × 8 mm on the bottom part

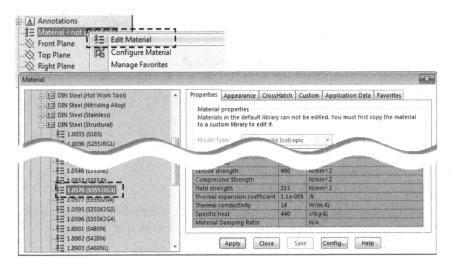

Fig. 5.29 Setting the material characteristics of the model

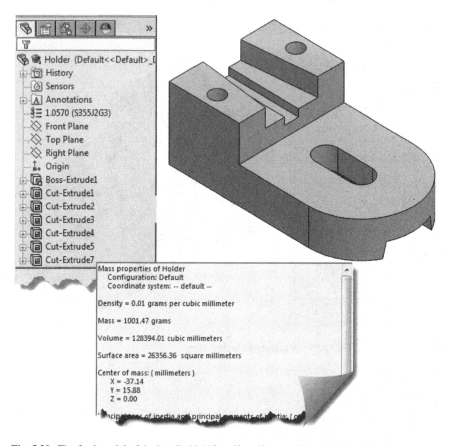

Fig. 5.30 The final model of the handhold $110 \times 60 \times 40$ mm and its characteristics

5.4 Modelling in NX

5.4.1 The Creation of a Model

We start the modelling with the command *New* in the *Start* menu. Next, we choose a working space and the template in which we wish to work (Fig. 5.31).

5.4.2 The Positioning of an Object in Space

The interface window and the *Working Space* opens on the screen. The coordinate system usually appears in the middle. To create the starting sketch we need *Datum Coordinate System* (Fig. 5.32), which represents the supplementary coordinate

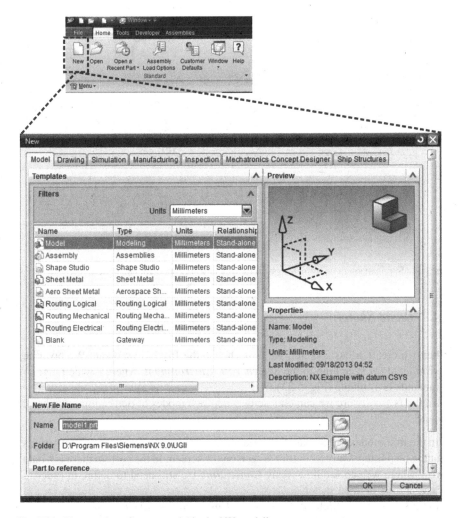

Fig. 5.31 The creation of a new model in the NX modeller

system. We can find the supplementary coordinate system in the ***Ribbon*** bar. In this coordinate system we choose between three planes and three axes and we can position them according to the natural product positions. The ***Working Coordinate System-WCS*** is usually positioned in the same ways as the supplementary coordinate system. When opening a new document, they are both positioned in the coordinate system of the model (0,0,0). We can rotate or translate the ***WCS*** in the space after we activate it in the dynamic shape. We do that with a double left-click of the mouse.

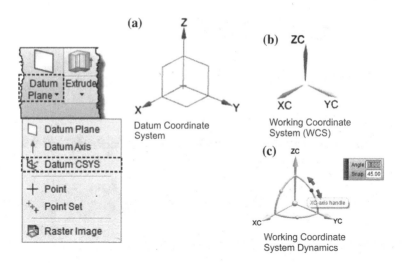

Fig. 5.32 Supplementary coordinate system (**a, b**) and working coordinate system (**c**)

5.4.3 Basic Sketch

When we create a basic sketch we first choose the front view plane. We open an environment for sketching in the **Sketch Task Environment**, where a wide range of commands is available (Fig. 5.33).

In the NX software we can choose between two modes of determining the basic sketch:

- In the first mode a basic sketch is without any defined dimensions and limitations at the beginning. Only a contour is seen on the screen. In the bottom there is a data row where we can see how many **Constraints** we still need to define in order to have a fully defined sketch.

- In the second mode (default) the function **Continuous Auto Dimensioning** is automatically switched on. It can be switched off in the submenu **Task Environment** (Fig. 5.34). The **Continuous Auto Dimensioning** function automatically shows the dimensions of the contour that we are drawing. But those dimensions do not define the sketch. For rapid use this mode is useful because the automatic dimensions define the rest of the undefined dimensions (in magenta). In any case, we need to fully define the sketch by ourselves. The automatic dimensions and the constraints level disappear and only the blue-coloured dimensions and symbols for the constraints remain. The symbol **p** and a serial number for the parameter appear next to the displayed dimension values on the dimensions line (Fig. 5.35). In use the undefined dimension means that the geometric entity can move freely, which is produced with the **MB1** click on it and a drag.

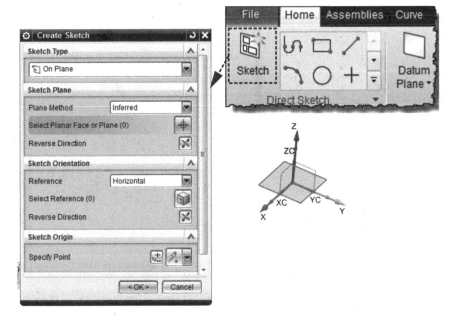

Fig. 5.33 The process of creating a basic sketch (first step, second step and third step)

Fig. 5.34 Activating the drawing of a basic sketch in the working space (*Sketch Task environment*)

5.4.4 The Creation of a Basic Sketch

The basic sketch is modelled with the feature *Profile* and is defined as shown in Fig. 5.35. To set the elements of a basic sketch we use *Constraints*, in our case *Tangent, Horizontal, Vertical* and *Perpendicular* (Fig. 5.36). The definition of the dimensions is concluded with two dimensions: the length of the basic line, 80 mm; and width of the basic line, 60 mm (Fig. 5.37).

The basic sketch is finished after we click on the window *Finish sketch* and activate the command for the extrusion in the *Ribbon* bar in the tab *Feature > Extrude* (Fig. 5.38).

We choose the basic sketch and set the length of the extrusion between the value of 0 mm and the final value of 40 mm. We orient the vector perpendicular on the plane of the basic sketch.

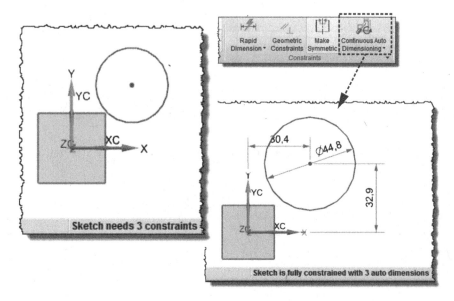

Fig. 5.35 Undefined basic sketch in the first mode (*left*) and automatically defined basic sketch in the second mode (*right*)

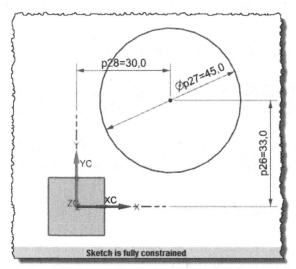

Fig. 5.36 Fully defined basic sketch (automatic dimensions disappears)

5.4.5 Cut-Out on the Front Part of the Model

To model a cut-out we need to draw a straight line using a 2D sketch on the top plane and we set its distance to be 40 mm from the basic plane, where the basic edge of the model is defined (Fig. 5.39). We finish the basic sketch with the gradual selection

List of all the available constraints

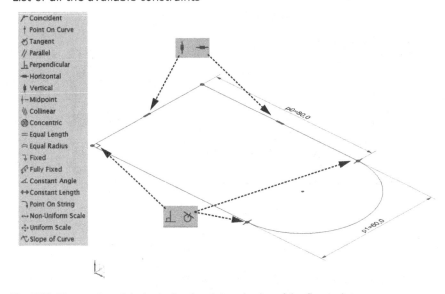

Fig. 5.37 The creation of the basic sketch and the selection of the *Constraints*

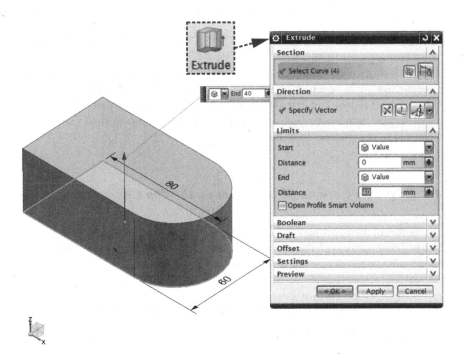

Fig. 5.38 Modelling the extrusion that is perpendicular to the basic sketch

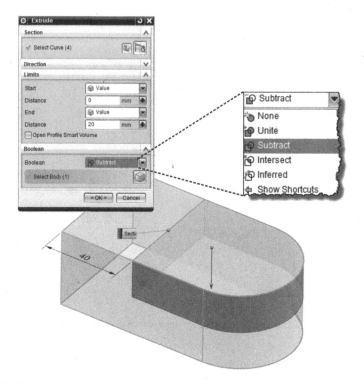

Fig. 5.39 The creation of the cut-out with the process of removing the material (*Subtract*)

of the edges (lines, circumferences, curves, etc) which close the area of the desired fragment. Now we have a new basic plane, which is not located in the plane of the first basic plane and we can extrude it to obtain the volume of the part of the model we wish to cut. In the same way we can optionally make other desired extrusions. We can add or cut volumes from the model.

5.4.6 Prismatic Cut-Out 30 × 10 mm on the Top Upper Part

In the middle of the upper part of the model we create a sketch of a four-sided prism, which is a cone shape with the dimensions from Fig. 5.40. We perform the extrusion perpendicular on the sketch plane with a deepening of 10 mm.

5.4.7 Modelling of a Small Groove 8 × 4 mm on the Bottom of the Cone Cut Out

We model a basic sketch with the total length and width of the groove on the already cut surface of the cone. Subsequently, we perform an extrusion with a depth of 4 mm (Fig. 5.41).

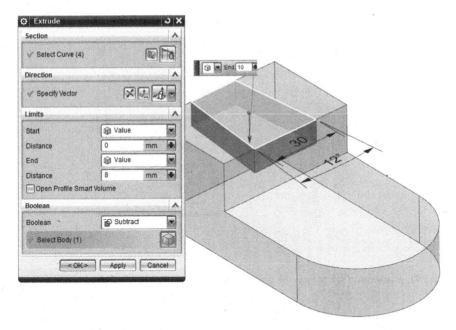

Fig. 5.40 Modelling of a four-sided prism in the shape of a cone for modelling an extrusion 30×10 mm on the top part of the model

5.4.8 Modelling of Through Holes $\phi 8.5$ mm on the Basic Sketch

We start the modelling by selecting the position of the initial coordinate point of the sketch so that it lies in the middle of the model width according to the upper surface of the model. The circumference $\phi 8.5$ mm is defined only once. Later we use the feature *Mirror curve* (Fig. 5.42), which mirrors the original circumference through the y axis of the coordinate origin of the current 2D sketch (Fig. 5.43).

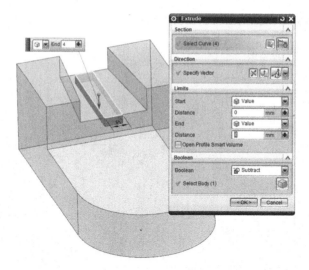

Fig. 5.41 Extrusion of the model with a groove of 8 × 4 mm, which is performed with the same features

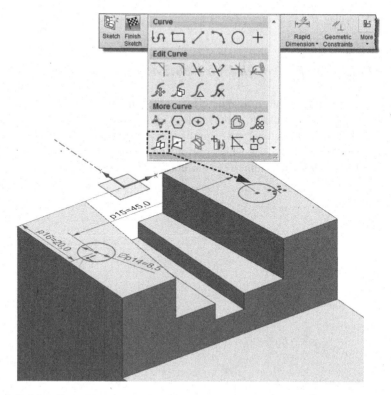

Fig. 5.42 Modelling of the holes $\phi 8.5$ mm on the upper surface of the model

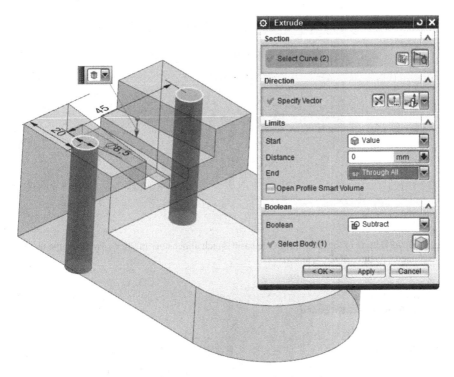

Fig. 5.43 Extrusion of the circumference for the production of holes $\phi 8.5$ mm through the entire model

5.4.9 Modelling of a Slot $\phi 13 \times 20$ mm

We begin the modelling with the activation of a new sketch (*Sketch*) and the positioning of the coordinate origin of the basic sketch in the centre of the main groove $\phi 13 \times 20$ mm (Fig. 5.44). We draw the basic groove sketch with the feature (*Sketch/Profile*) and we position the contour symmetrically to the local coordinate system of the sketch in this plane. We perform this with the feature (*Sketch/Make Symmetric*). After we conclude the sketching of the groove we finish the sketch (*Finnish Sketch*). We continue the modelling using *Feature > Extrude*, with which we create the extrusion of the recent sketch. To define the upper border of the extrusion we use the *Through all* function, for the bottom border the pre-set value of 0 mm (*Value*) remains (Fig. 5.45).

5.4.10 Modelling of a Special Groove 25×8 mm

Special groove can be made according to the Fig. 5.46.

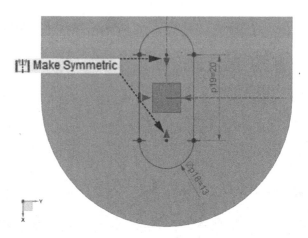

Fig. 5.44 Modelling of the groove with the basic sketch dimensions of 20 × 13 mm and the use of the symmetry feature (*Make Symmetric*)

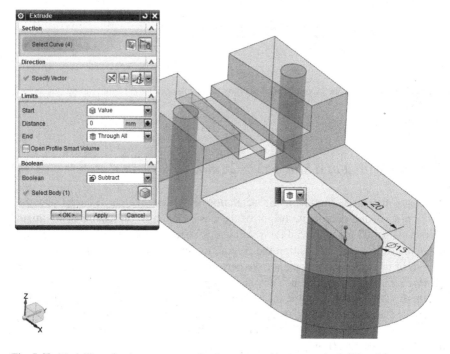

Fig. 5.45 Modelling of a slot or a cut-out for the groove with an extrusion of 13 × 20 mm

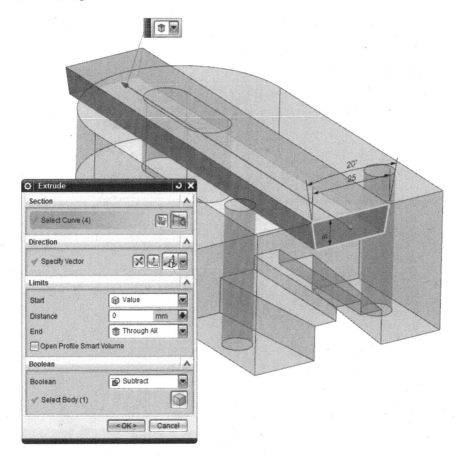

Fig. 5.46 Special shape of the groove on the bottom side with a profile with an inclination of $10°$ and dimensions 25×8 mm

5.4.11 Final Model

We can present the final model if we define its characteristic values, which we obtain by adding the material properties to the 3D geometry (*Tools > Utilities > Assign Materials*) (Fig. 5.47). Later, we analyze the characteristics of the model (*Analysis > Measure Bodies*). At the same time we can measure the mass, which the software calculates from the volume of the geometry, the static moment of the inertia and other geometrical characteristics. All the information can be displayed with the function activation of the tab *Show Information Window*.

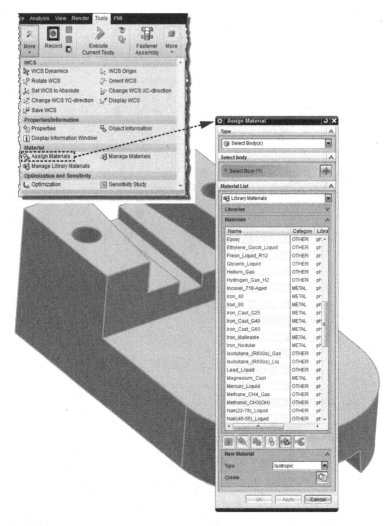

Fig. 5.47   To define the material of the 3D geometry we use the command ***Tools > Utilities > Assign Materials***

5.5 Examples

Examples of prismatic models are presented on Figs. 5.48, 5.49, 5.50, 5.51, 5.52 and 5.53.

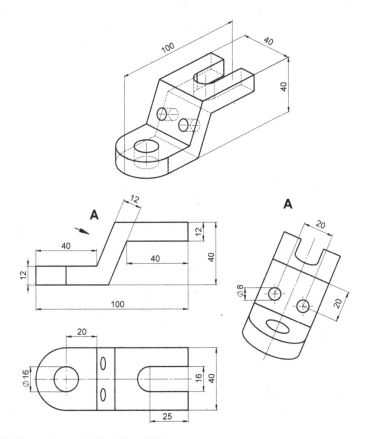

Fig. 5.48 Prismatic model 40 × 40 … 100 mm

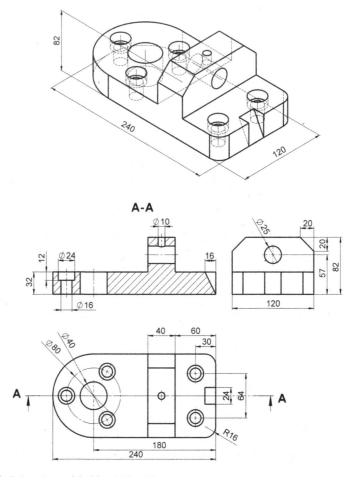

Fig. 5.49 Prismatic model $120 \times 240 \times 82$ mm

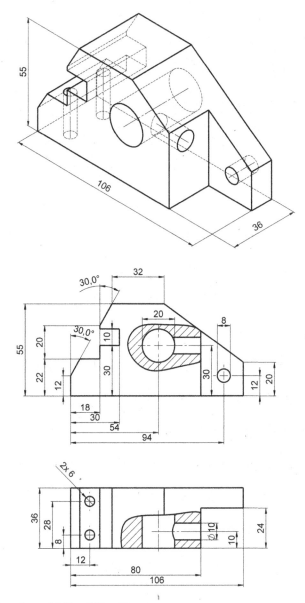

Fig. 5.50 Prismatic model 36 × 106 × 55 mm

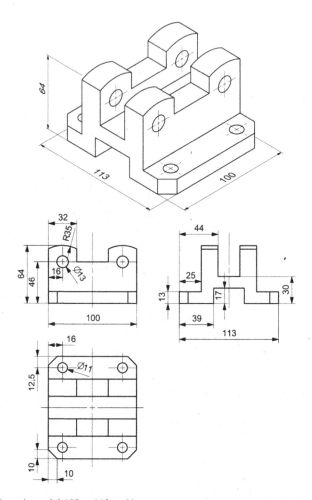

Fig. 5.51 Prismatic model $100 \times 113 \times 64\,\text{mm}$

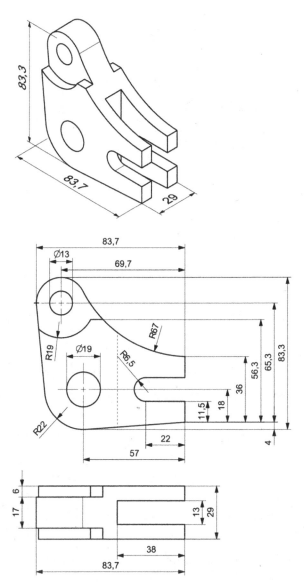

Fig. 5.52 Prismatic model 29 × 83.7 × 83.3 mm

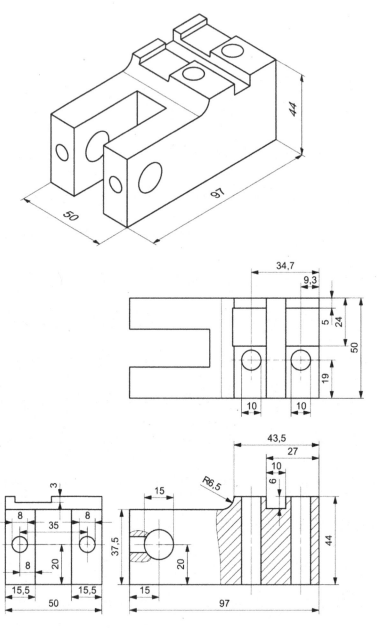

Fig. 5.53 Prismatic model 50 × 97 × 44 mm

Chapter 6
Revolving

Abstract Revolving is important for turned parts, spherical polishing and many other spherical forms (clay pots, round glasses, etc.). It begins with a base sketch. The revolving can either be fixed or it can be made open for additional features. Different approaches are presented, including mirroring across a specifically created plane. A shaft example is studied for different purposes, using a variety of important elements and shapes. Axisymmetric models that represent characteristic mechanical elements (e.g., wheel, shaft, pulley, belt pulley, bolt, etc.) are normally produced using a turning process. Here, we model them with a process of revolving. The starting point for modelling axisymmetric models is similar to the one for prismatic models. Here, we also begin by drawing a sketch of the cross-section contour (for example, a shaft), which we then rotate around its axis. The axis of rotation is usually the x axis of the selected coordinate system. Since this is an axisymmetric model, we can use just a half of the cross-section and revolve it around the desired axis to obtain the solid object. In engineering language we would say that the model is revolved around the centre line, which represents one of the local coordinate axes in the sketch. There are several ways to create this kind of model. In this chapter we will present the simplest method for a relatively complex shape. The exercises also provide different approaches for these complex shapes. Using the approaches made with SolidWorks and NX that are presented here the reader will easily be able to identify the individual steps.

6.1 Manufacturing Technology

Axisymmetric parts are generally produced by turning, rotary grinding, drilling, milling, etc.

In these approaches it is typical for a cutting point to rotate around one axis. The process of creating a model is similar to the one we use with prismatic models, the difference is that here we have a circle instead of a line. The most common process

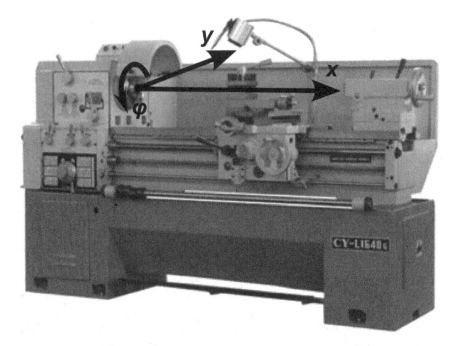

Fig. 6.1 3D space by turning with a defined axis. φ coordinate, representing 'z' axis, is defined as rotation around x axis. This is sometimes named 2.5D space

of turning is the one that is processed in the same way on the classic (Fig. 6.1) and the newer CNC machines. It is important that we understand the movement of the tool in relation to the part we are working on.

6.2 Modelling of Axisymmetric Models

We model axisymmetric models on the principle that the contour of a round object is rotated around the main axis or the axis of symmetry. In other words, this is a rotation contour in a sketch, which when rotated around an axis by 360° creates a surface and, consequently, a 3D object. The revolving procedure is analogous to that of extrusion, to the extent that the vector of the extrusion is substituted by an arc or a circumference. There are three important parameters:

- The sketch that includes the contour
- The main axis of rotation
- The angle of the sketch rotation about the main axis

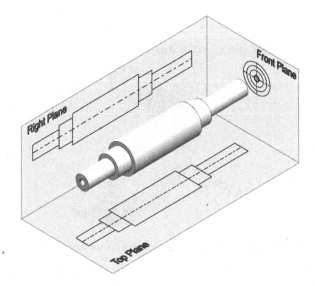

Fig. 6.2 Positioning of an object in space

6.2.1 Positioning of an Object in Space

The main characteristic when positioning an axisymmetric object in space is that we obtain the same object projection in both the top and front views (Fig. 6.2). In the side view, however, we usually have the projection of a circumference.

6.2.2 Positioning of the Basic Sketch

The position of the sketch in space is set by the selection of the planes and the axes of the coordinate system (CS). Usually, we define one of those axes as the axis of rotation (Fig. 6.3). The basic sketch always has to lie on the half plane with respect to the axis of rotation, which needs to be in the plane of the sketch (Fig. 6.4). Finishing the sketch is an important part. Some modellers require a reference line on the axis of rotation; others allow the usual contour or even just the axis of rotation, which is defined by the global coordinate system or a working coordinate system.

In general a sketch that does not lie on both half planes of the rotation axis can lead to an error during the revolving. The operation is not usually executed and the program reports an error.

In more advanced modellers the set of basic conditions is not limited only to the sketch drawn on the plane and the axis of rotation. The revolving of free-form curves is also possible (Fig. 6.5), as are more complex, if we use the Boolean operation (Fig. 6.6).

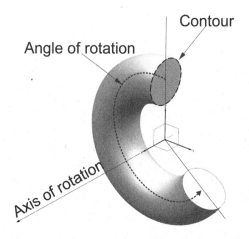

Fig. 6.3 Presentation of the revolving of a torus with the main information

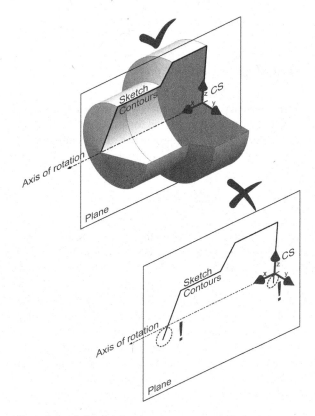

Fig. 6.4 Correctly performed sketch (*above*) and wrong sketch (*below*)

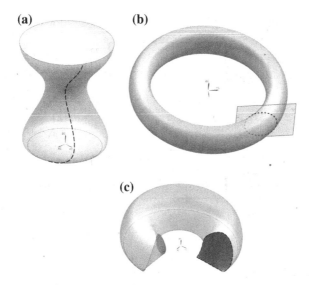

Fig. 6.5 Examples of the complex revolving of 3D space geometrical entities. 3D Spline (**a**), rotation axis out of sketch plane (**b**), revolved free face (**c**)

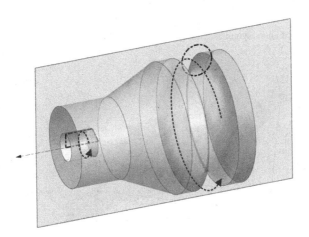

Fig. 6.6 Construction of more complex rotational bodies using the Boolean operation

Later on we will present the process of constructing the axisymmetric model of a shaft, which is shown in Fig. 6.7.

The main steps of the process are (Fig. 6.8):

1. Definition of the basic form of the shaft with the main cross-section contour.
2. Formation of the centre bore.
3. Formation of the retaining ring grooves.

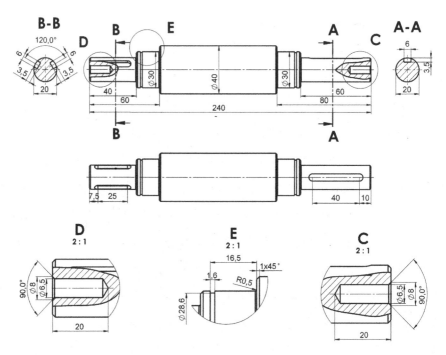

Fig. 6.7 Example of an axisymmetric model of a shaft ⌀40 × 240 mm

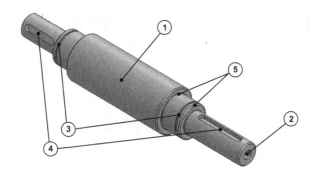

Fig. 6.8 Steps in the modelling process for an axisymmetric model of a shaft

4. Formation of the keyseat.
5. Improvements to the model regarding the specific manufacturing technology (edge filleting, chamfering, etc).

6.3 Modelling in SolidWorks

The start is similar to the modelling of the prismatic model. After the start-up of the program we begin by creating a new part (*File > New > Part*). We set the position of the model by the selection of a sketch plane. Since it is an axisymmetric model and the type of machining is determined (e.g., turning), the axis of the model should correlate with the rotation axis of the machine. In our case the rotation axis correlates with the x axis of the selected coordinate system. We draw the basic sketch (*Insert > Sketch*) on the front-view plane (x–y) (Fig. 6.9).

6.3.1 Basic Shaft Form

We produce the basic sketch of the cross-section or half cross-section on the selected plane. Usually, we start the sketching by drawing the centre line, which represents the axis of rotation and correlates with the x axis of the selected coordinate system. We position the starting point at the coordinate origin. The half cross-section contour we determine with respect to the shape of the model. The half cross-section contour is sketched in advance so as to be able to insert the values into the program more quickly. We produce the contour by following the relations between the individual entities (*Horizontal, Vertical*, etc).

The size of the object is defined by the dimensions. The dimensioning of the axisymmetric models is performed in the same way that the piece will be manufactured. The individual steps from the left and the right sides we dimension in parallel.

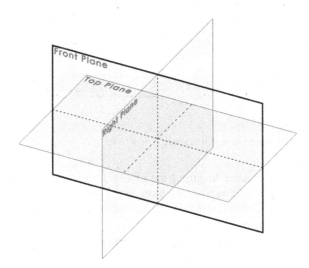

Fig. 6.9 Selection of the sketch plane in the case of an axisymmetric model

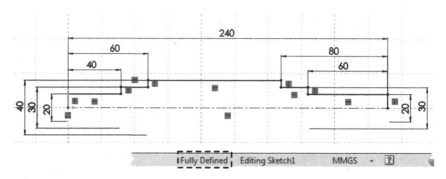

Fig. 6.10 Basic sketch of an axisymmetric model. The dimensioning of the diameters is a specialty in SolidWorks

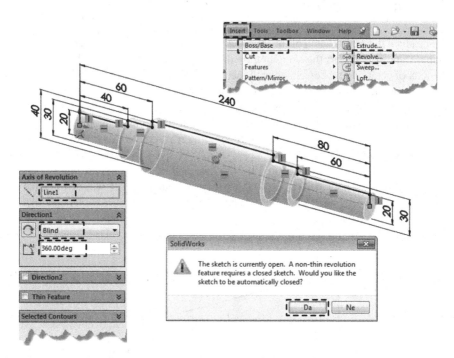

Fig. 6.11 Revolving of the basic sketch in space

We also dimension the entire length of the object. In the case of dimensioning a diameter we can help with dimensioning through the centre line (Fig. 6.10).

We get the basic form of the model when we revolve the sketch around the selected axis (***Insert > Boss/Base > Revolve*** …) (Fig. 6.11). It is important that the sketch is completely defined (***Fully Defined***). For the solid model the contour usually has to be closed, but SolidWorks can close the contour automatically in the case of revolving. First, we need to define the axis of rotation and the parameters of the revolving. In our case it is a blind revolve (***Blind***) in the full angle (360°).

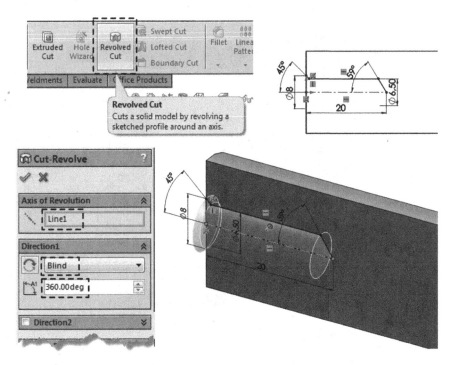

Fig. 6.12 Modelling of the centre bore on the *left side* of the shaft

6.3.2 Formation of Centre Bores

By fixing the elements on a shaft (pulley, gear, etc) against the axis movement, two blind screw holes are provided on the left and the right sides of the shaft. In relation to the size of the screw, we first need to make a hole with a suitable diameter. The process of creating the hole is similar to the process of making the basic form. But when we are adding the shape of the hole in the shaft we need to use the command for the removal of the material. We make the cut-out with the command *Insert > Cut > Revolve* or with a click on the icon *Revolved Cut* (Fig. 6.12). Later we select the sketch plane (Front Plane) and draw the shape of contour on it.

To make a centre hole on the opposite side we can use an identical process. However, it is easier if we mirror the already-made sketch of the model of the centre hole through the symmetric line (Fig. 6.13). We execute this in order to open the existing sketch (*Sketch 2*), and then create the centre line that connects both last points of the model. We draw the symmetric line in the middle (*Midpoint*), perpendicular to the axis, through which we produce the mirroring (*Mirror Entities*). Finally, we close the sketch (*Exit Sketch*) and restore the model (*Rebuild*).

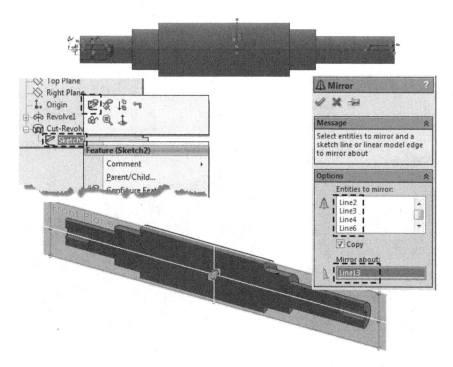

Fig. 6.13 Mirroring the sketch of the centre bores through the symmetry and cross-section of the model of the shaft with the centre bores

6.3.3 Formation of the Retaining Ring Grooves

We can draw the groove for the retaining ring while creating the base sketch or we can form it with an additional feature to remove the ring from the basic form (***Insert > Cut > Revolve***). The advantage of using the additional feature instead of forming the retaining ring groove in the base sketch is a possibility to freeze the feature, if needed. If the keyway slot is outside the specifications for the basic form, we can freeze it when we do not need it or it can also be used elsewhere. We can position the sketch of the shape of a retaining rings groove in any plane. In our case we position it in the same plane as the base sketch (plane x–y). We can adopt the axis of rotation from the base sketch or we draw it a new. Because both keyseats are the same we can add the connection between them. We dimension the size and the distance of the retaining rings groove from the wall (the width of the bearing and retaining ring together). We perform revolving with the command ***Revolved Cut***. We start in the sketch plane and produce a blind revolve on the perpendicular on the sketch plane for the full angle (360°) (Fig. 6.14).

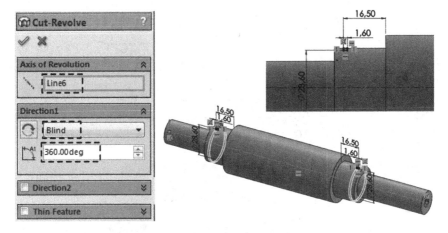

Fig. 6.14 Modelling the retaining ring groove and their positioning in the model of the shaft

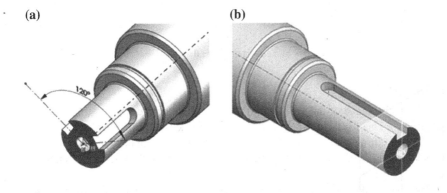

Fig. 6.15 Two keyseats on the side of the gear (**a**) and one keyseat on side of pulley (**b**)

6.3.4 Formation of the Keyseats

To ensure the functionality of the shaft for the torque moment transmission, we need to format a keyseat on the left and right sides. The torque moment transmission at the pulley ensures one key, which is the reason we have one slot. On the other side there is a torque moment transmission from the shaft on to the gear. The shaft hub by the gear is shorter than the shaft hub by the pulley and in that case we need two keys with it. Two keys by one shaft hub require arc angle of 120° between them (Fig. 6.15).

The keyseat is made with a small milling cutter. We add an auxiliary plane, which is tangential to the plane in which we wish to form the slot. This new planar sketch is

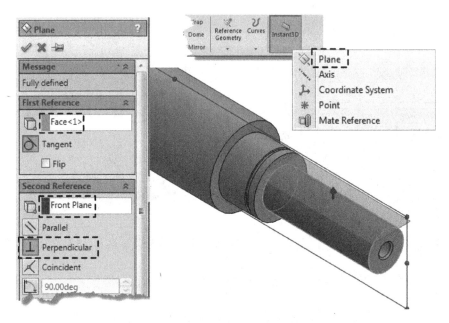

Fig. 6.16 Formation of the auxiliary plane at a tangent to the *cylindrical shape* and perpendicular to the front-view surface

the base for the shape of the keyseat and the same time defines the starting position for milling or else for the cut-out of the shape of the keyseat into the shaft.

We first create an auxiliary plane at a tangent to the surface of the diameter $\phi 20$ mm and perpendicular to the front-view plane (x–y) (*Insert > Reference Geometry > Plane*) (Fig. 6.16).

We draw the sketch of the keyseat in an auxiliary plane that is determined as a tangential plane to the cylinder, which represents a part of a shaft. To perform that we use the tool (*Straight Slot*). The size of the slot is 40×6 mm and it is 10 mm from the end of the shaft.

With the command *Extruded Cut* we extend the base sketch in space, in our case towards the centre of the shaft by the value of the slot depth, which is 3.5 mm (Fig. 6.17).

We have two keyseat on the side of the gear, which lean at an angle of 60° from the left and the right of the central, front plane (Fig. 6.15). We first need to form the auxiliary plane that goes through the axis of the shaft and is positioned at an angle of 60° to the front-view plane. Later we form another plane that is perpendicular and tangential at the same time to the cylindrical shape (Fig. 6.18).

The modelling of the keyseat on the side of the gear is similar to the one on the side of the pulley (Fig. 6.17). So we can draw a sketch of the shape of the slot on the already-made auxiliary plane and we extrude it into space (*Extruded Cut*). The diameter of the shaft is the same here, so the shape of the keyseat is also the same (the width of the slot is 6 mm, and the depth is 3.5 mm). The difference is in the

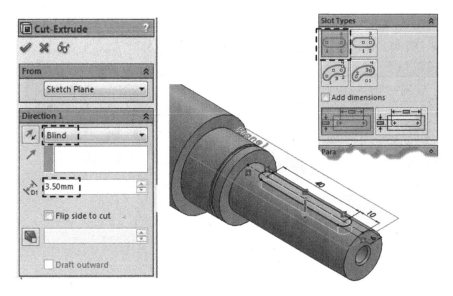

Fig. 6.17 Formation of the keyseat on the side of the pulley

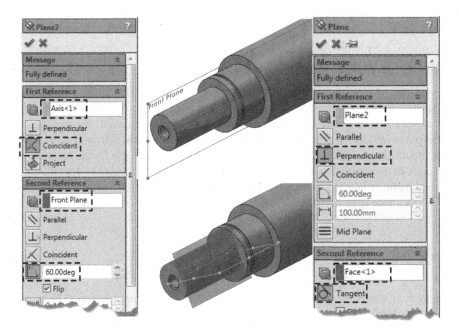

Fig. 6.18 The auxiliary plane that goes through the axis of the shaft and is positioned at an angle of 60° to the front-view plane and the second auxiliary plane, which is perpendicular to the first one and at the same time at a tangent to the cylindrical shape

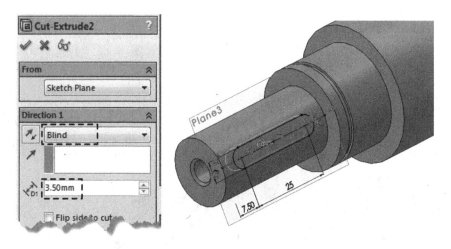

Fig. 6.19 Formatting of the keyseat on the side of the gear

length, which is less in the case of the gear-keyseat (25 mm). The slot is 7.5 mm from the greater diameter of the shaft (Fig. 6.19). The second slot, which is on the side of the gear at an angle of 120°, we produce in the same way.

6.3.5 Edge Filleting and Chamfering

A special feature is provided for modelling the edge filleting and chamfering. In the case of axisymmetric parts we can form them in the base sketch. We present an example to inform the user about different modelling approaches, from which he or she should always choose the one that is simplest and fastest to use for the modelling process. In our case we correct the base sketch (*Edit Sketch*). We add edge filleting R = 0.5 mm (*Fillet*) in the inner corners and chamfering 1/45° (**Chamfer**) in the corners outside (Fig. 6.20).

6.3.6 Final Model

Figure 6.21 shows the final model of the shaft with its structure of features (*Feature Manager Design Tree*) and mass characteristics for the material S355J2G3.

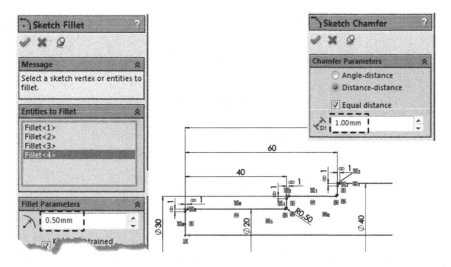

Fig. 6.20 Direct correction of the basic sketch—adding the edge filleting and the chamfering

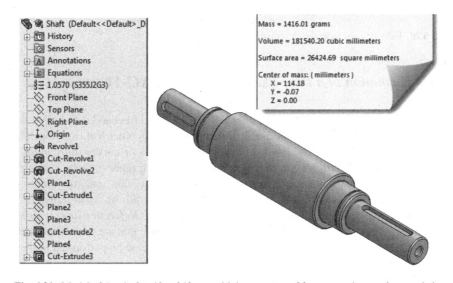

Fig. 6.21 Model of the shaft ∅40 × 240 mm with its structure of features and mass characteristics

6.4 Modelling in NX

6.4.1 Positioning of an Object in Space

We position an axisymmetric object (rotational bodies) in space and set the x axis as the rotation axis. At the same time we select the plane XC–YC for drawing the base sketch (Fig. 6.22).

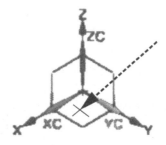

Fig. 6.22 Selecting the plane for the base sketch

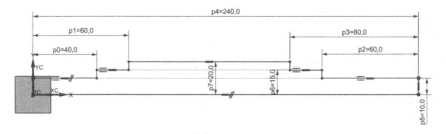

Fig. 6.23 Fully defined base sketch with dimensions and constraints

6.4.2 Formation of a Base Sketch in the Plane XC–YC

The base sketch has all the essential dimensions defined in a freehand sketch. When we input the freehand sketch we need to determine the geometrical, topological elements for 2D, defining the lines, arches, circumferences or curves (*horizontal, vertical, collinear, equal length, parallel*, etc.). The sketch represents the contour of the round object in half view. We rotate the contour by 360°, and so we close the volume and we obtain a 3D object. The NX modeller allows us to transform the contour or its part on to the auxiliary line for drawing (*Reference line*) with a right-mouse click and placing the cursor on the optional line that we select from the menu (*Convert to reference*) (Fig. 6.23). The process can also be performed in the opposite direction. The NX modeller will also complete the operation of the rotation around the axis if the axis of rotation is not a closed contour.

6.4.3 Revolving into Space

The process of 3D object creation is simple. We activate the feature for revolving in the main menu (*Feature > Revolve*). In the submenu (*Section*) we select the curve for the revolving. Later, with the command *Axis*, we select the vector or the axis and the origin of the arc that represents the curve of revolving (circumference). Following

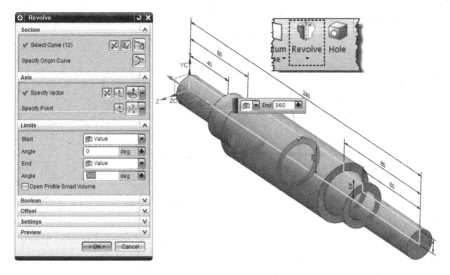

Fig. 6.24 Rotating bodies of the base sketch around the x axis by 360° enables the presentation of a 3D model

this we set the size of the rotation (the size of the revolving) with the command *Limits*, which is 360° in our case (Fig. 6.24).

We need to add some details to the base sketch because it has sharp edges. In this phase of the modelling the details increase the database, but if the details are important for the function of the product it is appropriate to include them in the base sketch. The user has to decide about the importance of the details during this phase. If we want to model the details now we need to switch off the *Preview*. This allows us to correct the base sketch (Fig. 6.25).

In the base sketch we fillet the inner edges with radius R = 0.5 mm and chamfer the sharp edges at a length of 1 mm for an angle of 45°. To do this we use the features *Sketch/Fillet* and *Sketch/Chamfer* (Fig. 6.26).

6.4.4 Modelling a Centre Bore

We model the centre bores in a similar way as we model the original basic sketch. We set the screen view in such a way that we are able to see clearly the whole model or part of the model that we are going to implement. We select *View > Visibility > Style* in the main toolbar to set the desired view. We select one of the shaft (the left or the right) and the crossing plane XC–YC. Then we draw a new, partial base sketch for the centre bore into the enlarged view (Fig. 6.27).

To model the centre bore at the other end of the shaft we can use the 2D feature for mirroring (*Mirror Curve*). Before the mirroring process we always have to set the

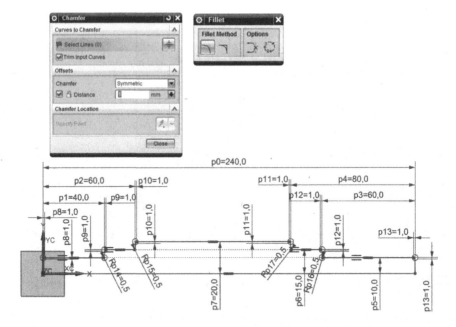

Fig. 6.25 Chamfer and fillet modelling on the base sketch

Fig. 6.26 Model of the shaft with chamfered sharp outer edges and fillet transitions from smaller to larger diameters

mirror line. To do this we mark the very left and very right points on the symmetrical axis of the rotational bodies, to which we later define the middle point with the activation of the automatic constraints (*Top Border Bar > Midpoint*). In this location we create a perpendicular mirroring axis, which we change into the reference axis (*MB2 > Convert to Reference*). The references can be of various distances. Finally, we conclude the sketch of both centre bores and activate the command *Revolve* for the full solid model (Fig. 6.28).

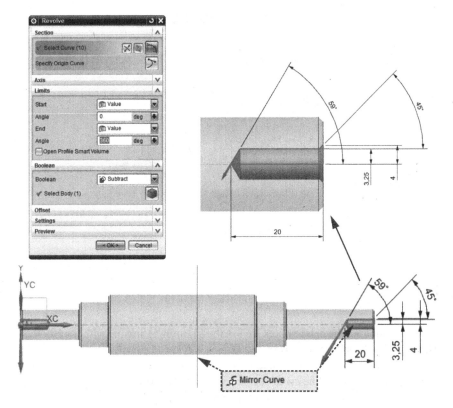

Fig. 6.27   Modelling of the centre bore ⌀6.5 mm

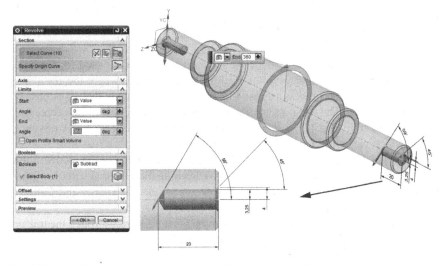

Fig. 6.28   Modelling of two centre bores by rotating around the main x axis and mirroring

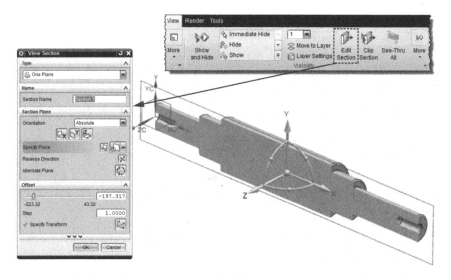

Fig. 6.29 View of the cross-section of the shaft in the plane XC–YC

When modelling the centre bores we use the trimming principle with the Boolean command (***Boolean/Subtract***) (Fig. 6.29).

It is advisable to view the whole object after every activity. In this modelling stage we use the view of the cross-section of the model, which we perform with the commands: ***View > Edit Section/Clip Section***.

6.4.5 Modelling of the Retaining Ring Groove

As already mentioned, we usually derive from the plane that we used for the base sketch. That is why we always come back to the base plane where the revolving was performed. In the process we use those geometric elements that most clearly represent the shape. The width and the depth of the retaining ring groove we represent with the a rectangle (Fig. 6.31). If needed, we draw the auxiliary geometry (reference line on which we connected the sketch of the groove).

Figure 6.30 shows presentation of the projection of the edge of the model in the sketch using the 2D feature ***Project Curve***. We again mark the curve with the right-mouse button (***MB2***). Then we activate the menu and select the command ***Convert to Reference***.

Using the auxiliary geometry we sketch groove that we mirror through the mirror line to the other side. The auxiliary geometry needs to be defined on the base of the previously set geometry. In the operation of revolving the retaining ring groove we perform the Boolean operation for trimming the material (***Revolve/Boolean > Subtract***).

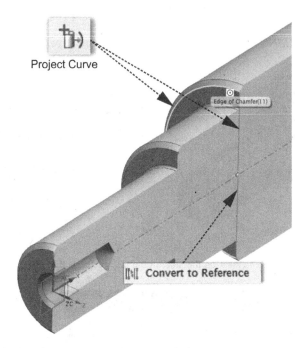

Fig. 6.30 Projection of the edge on the sketch plane laying

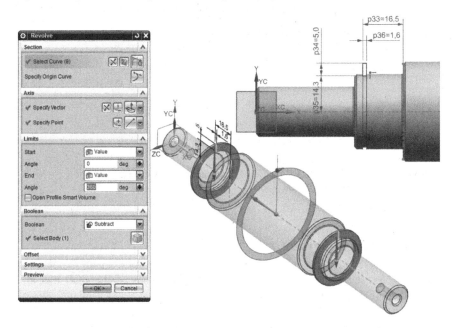

Fig. 6.31 Modelling the sketch and revolving the retaining ring groove

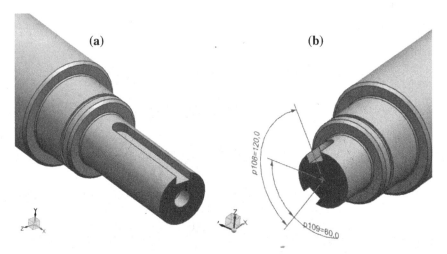

Fig. 6.32 Modelling of a keyseat on the side on the side of the pulley (**a**), on the side of the gear (**b**)

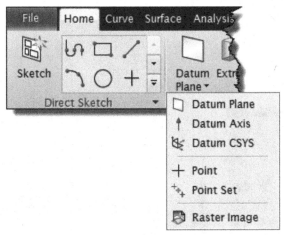

Fig. 6.33 Presentation of the menu for the auxiliary geometry that we use to create the axis, planes, supplementary coordinate system and points

6.4.6 Modelling a Keyseat

Figure 6.32 presents the keyseat on the left and right ends of the shaft that are modelled using the extrusion and trimming.

In the following we create the slot on the side of the pulley using the auxiliary plane (***Datum Plane***) (Fig. 6.33). We position the auxiliary plane tangentially to the cylindrical shape and rotate it perpendicular to the front-view plane (x–y) (Fig. 6.34).

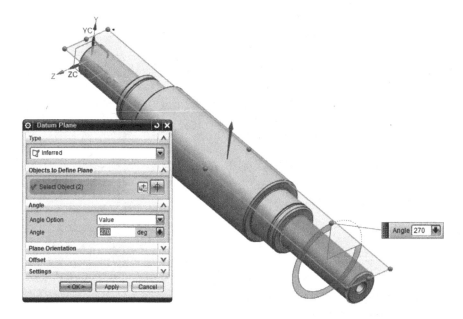

Fig. 6.34 Positioning of the plane for the extrusion that we use for the keyseat groove

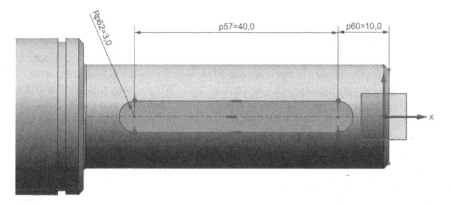

Fig. 6.35 Base sketch and dimensions for the keyseat 40 × 6 mm

To model the slot (Fig. 6.35) into the shaft with a depth of 3.5 mm for the keyseat we use linear extrude to cut-out the material. After we select the option **Boolean** we trim the material (**Subtract**). We define the vector of the extrusion using the command **Specify Vector**.

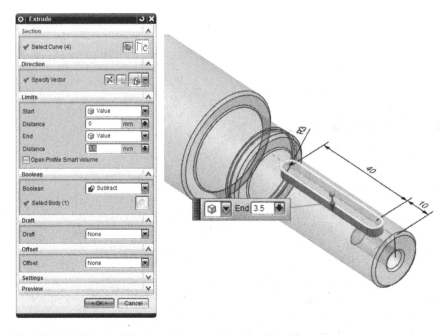

Fig. 6.36 Modelling the 40 × 6 mm keyseat with on the side of the pulley

When we model two keyseats it is recommended to represent the shaft as shown in Figs. 6.32b or 6.39. We position the auxiliary plane at a tangent to the cylindrical shape at a specific angle (in our case the angle is 60°) on the left side because of the following mirroring through the vertical axis. We perform this so that when positioning the plane under the option *Type* we select *Inferred* and two entities: the cylindrical shape and the plane XC–ZC (Fig. 6.36). The *Inferred* setting establishes the type of definition on its own, taking into account the user selection and the type of geometrical entities (Fig. 6.36). In the end we set the angle to 60° (Fig. 6.37), which determines the size of the half angle for the mirroring of the keyseat between the new plane and the plane XC–ZC.

After we position the plane (x–z) we draw a sketch to form the slot 25 × 6 mm (Fig. 6.38). We proceed with the extrusion to a depth of 3.5 mm in the direction of the centre of the shaft. On the side of the gear we have two keyseats, which are rotated by an angle of 120°. For the other two keyseats we draw a sketch on a plane, which we select in a similar way (Fig. 6.39). In this case the angle of the plane is 300° (Fig. 6.39). Final model of the shaft is presented on the Fig. 6.40.

Hint for additional options: in the case that we would like to find a specific feature or function, we can use the window for a search called *Command Finder*.

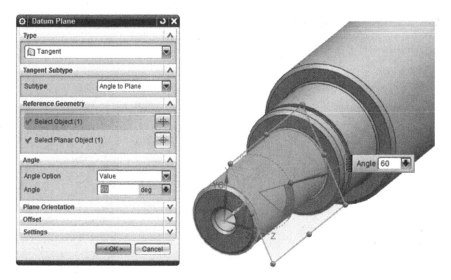

Fig. 6.37 Positioning of the auxiliary plane at a tangent to the selected plane using the optional angle regarding the front view plane (x–z)

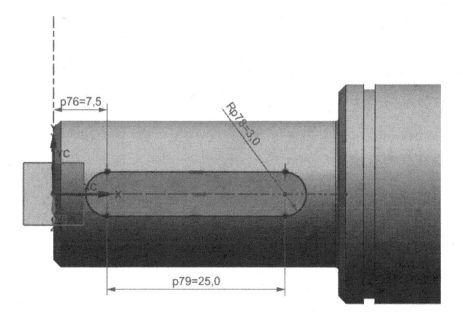

Fig. 6.38 Sketch of the keyseat 25 × 6 mm on the side of the gear

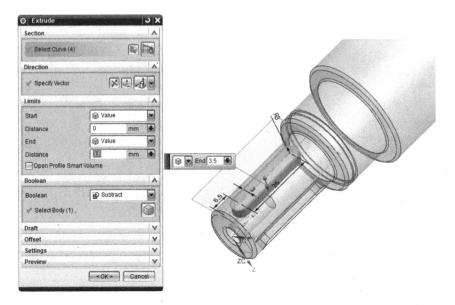

Fig. 6.39 Modelling of the slot for the extrusion to a depth of 3.5 mm towards the centre of the shaft

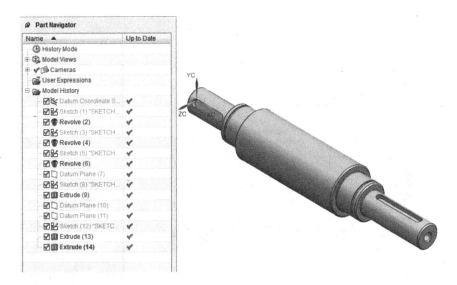

Fig. 6.40 Final model with the history of used features

6.5 Examples

Figures 6.41, 6.42, 6.43, 6.44, 6.45 and 6.46 present a few models that can be modelled to test the modelling knowledge.

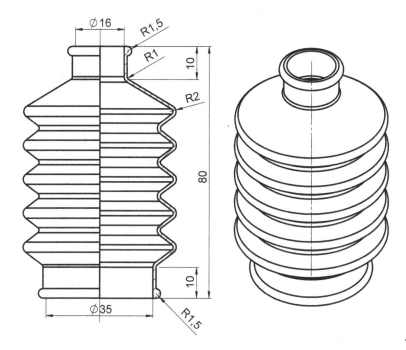

Fig. 6.41 Seal ⌀35/Ø16 × 80 mm

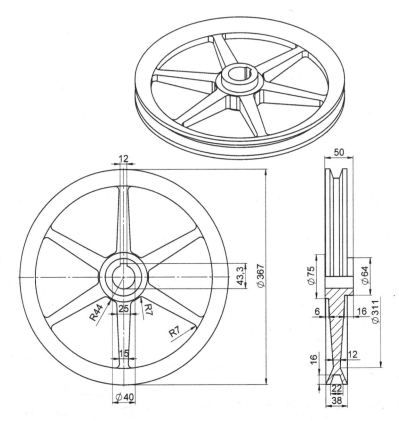

Fig. 6.42 Pulley ⌀376/⌀40 × 50 mm

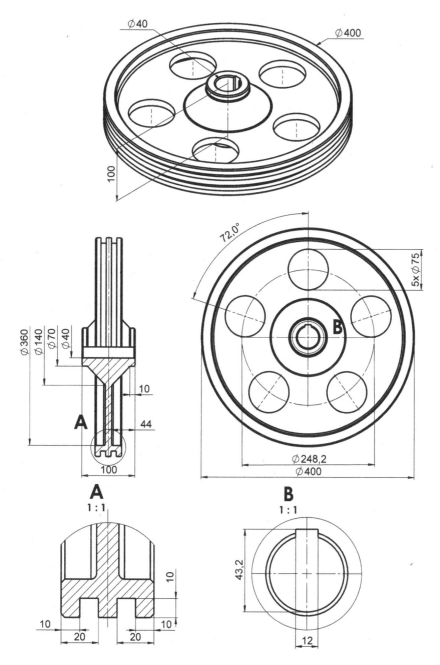

All fillets are R=3mm, all chamfers are 2 x 45°

Fig. 6.43 Friction wheel ⌀400/⌀40 × 100 mm

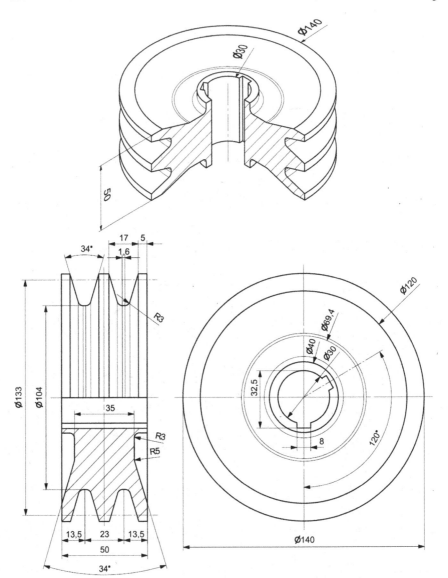

Fig. 6.44 Pulley $\varnothing 140/\varnothing 30 \times 50$ mm

Fig. 6.45 Hand griff
⌀83.6 × 45.6 mm

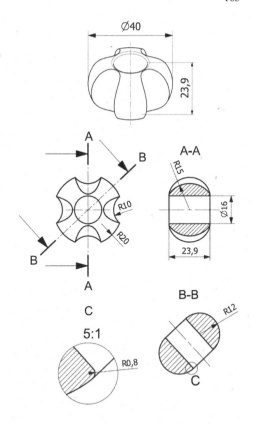

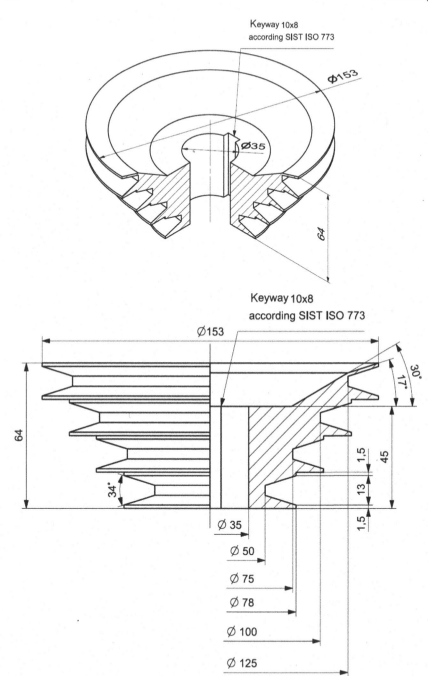

Fig. 6.46 Pulley ⌀153/⌀35 × 45 mm

Chapter 7
Sweep

Abstract Curve sweep represents an important general form for defining the spatial shapes of different cross-sections, changing along a pre-defined curve. This feature is particularly useful for creating pipelines and other free-form linear profiles. It is applied to different examples, such as springs or lighting equipment, which proves its universal applicability. Examples of complex shapes show possible applications in different fields of industry.

Curve sweep is important for creating objects that are defined at particular points by cross-sections. The connecting curve along the middle of these cross-sections is defined by a spatial curve. This allows for the creation of more complex designs with the constant cross-sections that exist in nature. A spring is a typical example of curve sweep.

This chapter will present the feature of curve sweep, including both constant and variable pitch. In addition to the above example, curve sweep is also used for pipelines (water-distribution systems, heating pipelines, air-conditioning channels, oil-hydraulics networks etc.), electrical cable installations, etc.

The advantage of curve-sweep modelling is particularly obvious when planning tubular installations for air-conditioning devices, electric installations (for ships and cars, for example). Tube installation modelling yielded neat installations in fixing installations, as well as a significant advantage in designing lengths.

7.1 Manufacturing Technology

Springs, as simple helices, are basically identical round cross-sections, coiled on a spiral curve of the same diameter. Spring-manufacturing machines therefore use a specific coiling angle, a system of rollers or sliders, which provide the plastic deformation of the wire, usually directly before the coiling. With deformations that are too large, it is necessary to initiate the rough shape prior to the second coiling in order to avoid too many residual stresses. Figure 7.1 shows an example of a spring-manufacturing machine.

J. Duhovnik et al., *Space Modeling with SolidWorks and NX*,   167
DOI: 10.1007/978-3-319-03862-9_7, © Springer International Publishing Switzerland 2015

Fig. 7.1 Spring-manufacturing machine

Fig. 7.2 The use of curve sweep for the high-pressure delivery of diesel fuel to an engine significantly improves the conformity and simplicity of the structure for attaching the tube

A spring can be coiled with the same diameter along its entire length; this results
in a helical spring. However, with a reducing diameter for a certain function, we
obtain a conical helical spring.

The curve-sweep principle is applied in industrial technology or, for example,
in the high-pressure delivery system of diesel fuel injection for internal combustion
engines, like the example shown in Fig. 7.2. The appearance of a tube installation,
performed according to the "mount-and-connect" principle, is shown in Fig. 7.3.

Fig. 7.3 Installation in an old car, which was performed according to the "mount (all modules) and then connect" principle (without a blueprint for the tubular structures)

Fig. 7.4 Placing an object in space and the projections on the respective orthogonal planes

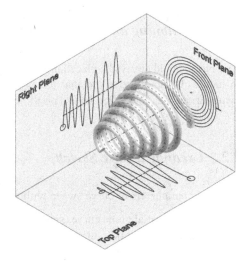

7.2 Modelling Products with Constant Cross-Sections

Modelling 3D objects with constant cross-sections is a concept analogous to the examples of extrusion and revolving. This feature also requires selecting a sketch plane with an open or closed contour. The guide curve is then selected, along with a geometrical shape (a circle, a ring, a square, a rectangle). By travelling along the guide curve we end up with a volume body or surface.

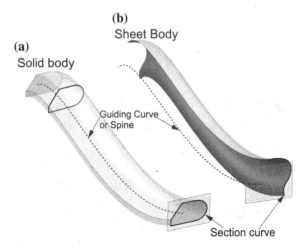

Fig. 7.5 Curve sweep with a constant cross-section for a closed (**a**) and open contour (**b**)

7.2.1 Positioning an Object in Space

An object should always be placed in space in such a way that it matches its position in nature. It should be noted that the objects created by curve sweep are rarely symmetrical, i.e., they often have a non-linear shape (Fig. 7.4).

7.2.2 Creating a Base Sketch

The basic parameters of curve sweep with a constant cross-section are (Fig. 7.5):

- A planar or spatial guide curve (spiral, B-splines, a set of connected straight lines)
- A planar or spatial guide curve (lines, an assembly of connected lines, arcs, splines, etc.)

7.2.3 Sweep

Setting a sketch plane that will be used to determine the base sketch is of key importance for determining a profile (Fig. 7.6). If the normal of the plane is tangential to the curve, this property is maintained throughout the sweep (Fig. 7.6a). However, if a plane is defined as being perpendicular to the curve, i.e., perpendicular to the tangent at a point on the curve (Fig. 7.6b), the cross-section along the curve remains constant.

(a)

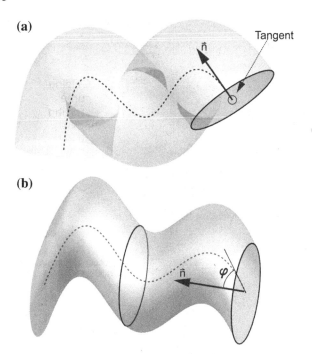

(b)

Fig. 7.6 Position of the base sketch plane relative to the position of the guide curve: tangentially plane (**a**) and vertical plane (**b**)

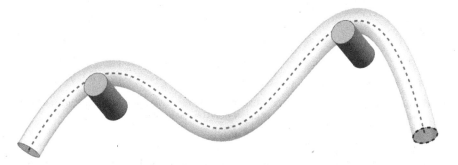

Fig. 7.7 Ratio between the size of a contour and the curvature of the guide curve

When modelling real products with curve sweep, it is necessary to take account of the minimum curvature of the guide curve in relation to the size of the diameter in the sketch. Otherwise, one surface will eat into the other that is coming out of the curve. An example is shown in Fig. 7.7.

With the latest modellers it is possible to perform curve sweep—a feature that does not require a strict connection between the guide curve and the base sketch plane (Fig. 7.8).

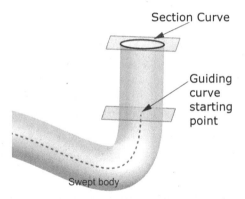

Fig. 7.8 The base sketch plane does not touch the guide line

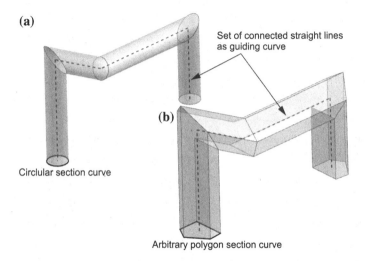

Fig. 7.9 An example of curve sweeps for two different profiles along a discontinuous guide curve—the skeleton. Circular section curve (**a**), arbitrary polygon section curve (**b**)

Curve sweep with a constant cross-section can also be performed along a polylines defined with many straight lines, generally referred to as the skeleton (Fig. 7.9).

What follows is an introduction to curve-sweep modelling for machine elements, i.e., a spring. In mechanical engineering, a flexible helical spring is the most frequently used type. The modelling of such a spring involves curve sweep.

The shape, created in a 3D space, requires at least two sketches in order to be able to define all three coordinates. For a spring, one sketch is used to define the profile, used for springs (it is usually the round cross-section of a round wire), and the other sketch is required to define the helix in space.

The manufacturing procedure for a helical spring model is presented in Fig. 7.10. The modelling procedure can be condensed into the following steps:

Fig. 7.10 A helical spring $\phi 100/\phi 10 \times 500$ mm

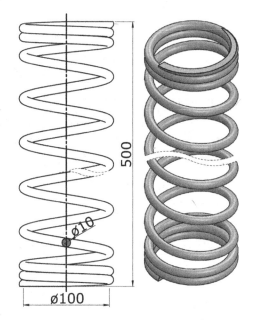

1. Creating a guide curve
2. Specifying a cross-section profile
3. Curve sweep
4. Shape complementing
5. Final model.

7.3 Modelling in SolidWorks

7.3.1 Creating a Guide Curve

When creating a helical spring model (Fig. 7.10), it is first necessary to create a guide curve in space, representing the path along which the desired profile will be extruded. In the case of a helical spring, this is a helix. It is created by extruding the base circle using the (***Helix/Spiral***) command.

The positioning of the base circle to form a spring and then the whole model in space is performed on a top view plane (Fig. 7.11).

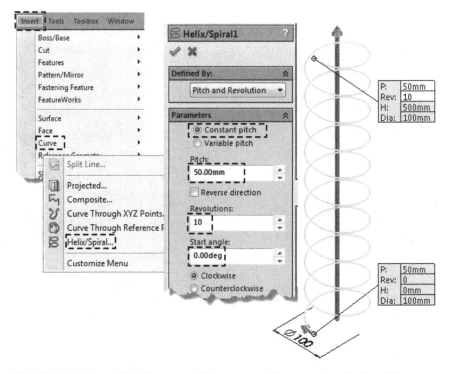

Fig. 7.11 Modelling a helix (diameter = 100 mm, step = 50 mm, number of coils = 10)

7.3.2 Setting a Cross-Section Profile

A cross-section sketch that is formed along the curve is defined on a plane that is perpendicular to the path curve. For this purpose, a reference plane is created. The plane is perpendicular to the extrusion curve and runs through the base point (***Insert > Reference Geometry > Plane***) (Fig. 7.12).

Draw a circle with a diameter of 10 mm onto the reference plane, with the helix piercing the circle right through its centre (***Add Relations > Pierce***) (Fig. 7.13).

7.3.3 Curve Sweep

Make a spring model by means of the ***Sweep*** feature. For the sweep, choose a pre-created profile and path, and then specify the other parameters that affect the extrusion (Fig. 7.14).

For a suitable load transfer, create the starting and ending coils at both ends. This should be done using ***Variable pitch***. This is achieved by correcting the helix-creation

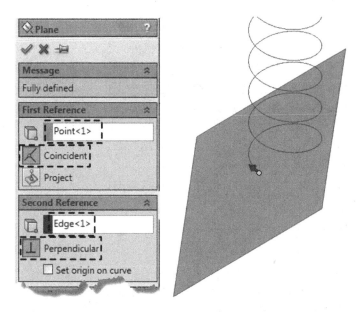

Fig. 7.12 Modelling an auxiliary plane perpendicular to the curve of the sweep

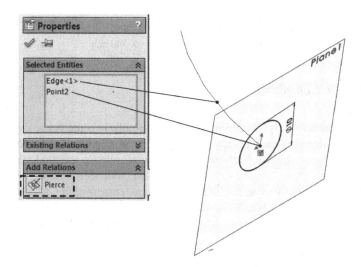

Fig. 7.13 Curve-sweep profile

feature (***Helix/Spiral***). Next, select a variable helix pitch instead of a constant pitch. Then define the helix pitch, the number of coils, the height and diameter, and the input data in the form of a table (Fig. 7.15).

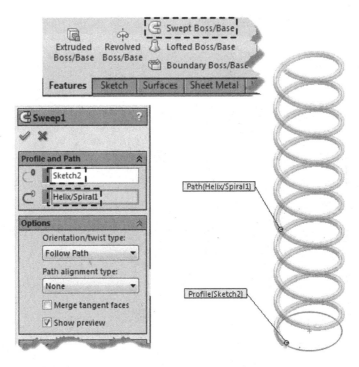

Fig. 7.14 A model of a simple helical spring $\phi 100/10$ mm $\times$ 500 mm

7.3.4 Shape Complementing

A compression helical spring absorbs the force evenly into the coils, making the upper and the lower planes evenly aligned. The coil alignment is performed by planar grinding, and the *Cut Extrude* function is performed in the modeller.

On a side-view plane, create a sketch of the cut-off cross-section. Find the shape of the existing geometry by using the intersection curve (*Intersection Curve*) as design lines (*For construction*) (Fig. 7.16). Create a sketch of the cut-off cross-section in the form of a rectangle and connect it to the existing geometry by means of relations. The extrude is normally performed on the sketch plane through the whole model (*Through All*) (Fig. 7.17).

7.3.5 Final Model

Figure 7.18 shows the finished model of a helical spring model and the corresponding structure of the features and the mass characteristics.

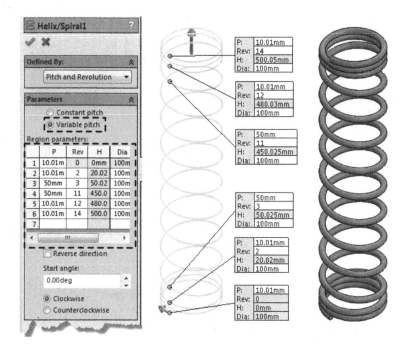

Fig. 7.15 Defining a helical spring with a variable pitch at both ends

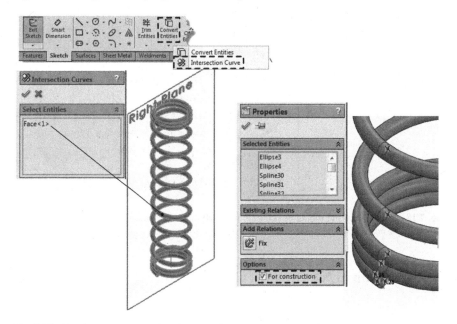

Fig. 7.16 Creating support polylines for modelling a sketch for cut-off on the top and bottom parts

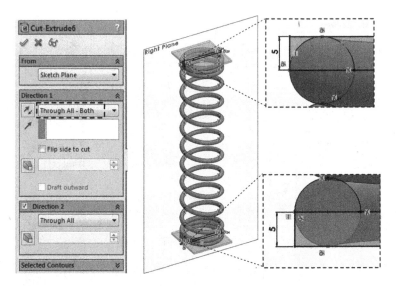

Fig. 7.17 Block extrude for a planar alignment at both ends of the spring

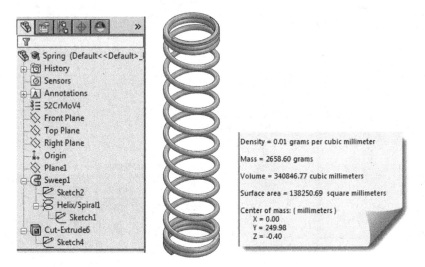

Fig. 7.18 A helical spring model $\phi 100/10$ mm $\times$ 500 mm with the finished starting and ending coils, and the corresponding mass characteristics.

7.4 Modelling in NX

Curve sweep begins by modelling the guide curve, which is then attributed to the shape of the cross-section. The details and the ending transitions are modelled at the very end.

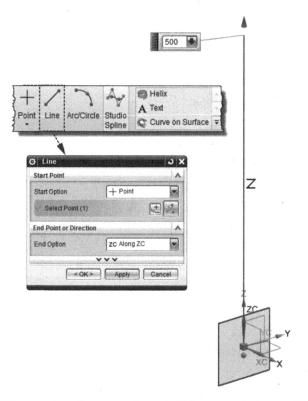

Fig. 7.19 Modelling a skeleton line for a spring in the Z direction and a length of 500 mm

7.4.1 Creating the Guide Curve

The guide curve, also referred to as the skeleton, is modelled along the whole length of the spring. It points in the z-axis direction (Fig. 7.19). For modelling, use a spatial line (*Curve/Line*). The skeleton line is required because the spring will have a variable coil pitch. To create the skeleton line, activate the (*Curve/Helix*) feature. Use a global coordinate system as the origin of the helix.

The (*Helix*) feature offers spring-radius options (*Size/Radius*), 50 mm in our case (Fig. 7.20). Due to the variable pitch in this example, the variable-pitch option will be selected in settings (*Helix/Pitch/Linear along spine*). In the (*Helix/Select spine*) tab, select a straight line for the spine (Fig. 7.21). Once the parameters have been set, proceed to creating individual points along the spine, where the location relative to the spine length and pitch per coil are defined. Select the points in the (*Specify new location*) tab and define them according to the scheme (Fig. 7.22).

Using the (*Helix/Length*) command, set the end distance to 500 mm. In the (*Helix/Settings*) tab, select the left-hand direction (Fig. 7.21).

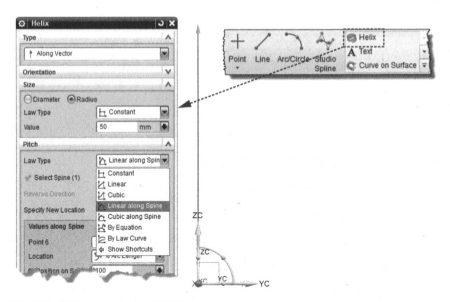

Fig. 7.20 Helix creation menu (*Helix*)

7.4.2 Creating a Cross-Section Profile

Having designed the helix, the profile needs to be modelled. In our case this is a round profile, representing a wire. Once the profile has been determined, extrude it by means of curve sweep to obtain the formed 3D object. The modelling with curve sweep is performed with the (*Feature/Tube*) feature. It allows the creation of a tube or a solid profile along a pre-defined curve, which is a helix in our case. For the profile, set the outer diameter at 10 mm and the inner diameter at 0 mm (Fig. 7.23).

7.4.3 Complementing the Technological Shape

During the final modelling phase, create the end of the spring. Do this by means of auxiliary planes (*Feature/Datum plane*). Select the plane by clicking the x-y plane of the coordinate origin and the point at the centre of the spring cross-section. For the plane type select (*Inferred*) (Fig. 7.24). Do the same for the other end of the spring.

Once the auxiliary planes have been set in their positions (Fig. 7.25), the rest of the material can be trimmed with the (*Feature/Trim Body*) feature. The trimming is performed by first choosing a body, followed by the plane with which to trim the body. In the (*Trim body/Reverse direction*) tab choose the direction of material trimming.

For a complete presentation of the model, inspect the process of creating the model (Fig. 7.26).

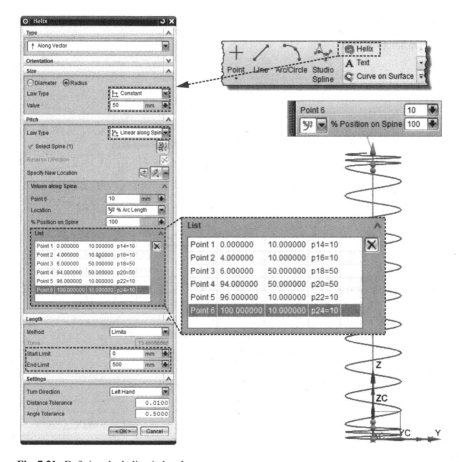

Fig. 7.21 Defining the helix pitch values

Fig. 7.22 Points defining
scheme for a variable pitch
along the spine

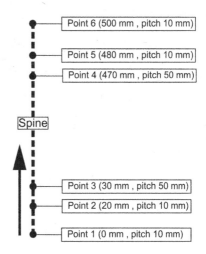

Fig. 7.23 Extruding a round cross-section with a diameter of 10 mm along the helix

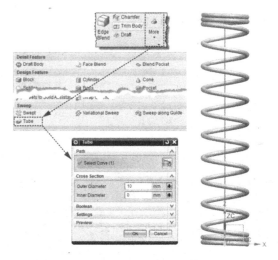

Fig. 7.24 Setting an auxiliary plane to create the ending

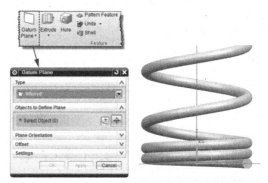

Fig. 7.25 Material trimming by means of an auxiliary plane

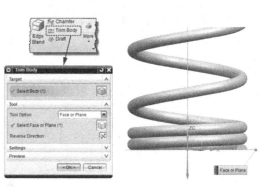

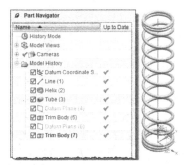

Fig. 7.26 A finished spring model $\phi 100/\phi 10$ mm $\times$ 500 mm with finished starting and ending coils and final chamfering

7.5 Examples

Figures 7.27, 7.28, 7.29, 7.30, 7.31, 7.32 and 7.33 presents few models od curve sweep parts.

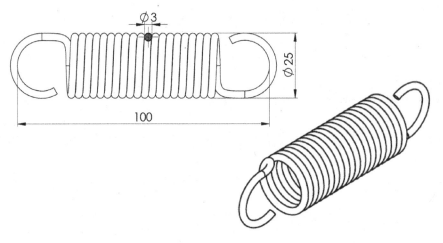

Fig. 7.27 Tension spring $\phi 25 \times 3 \times 100$ mm ($n_{ef} = 20$)

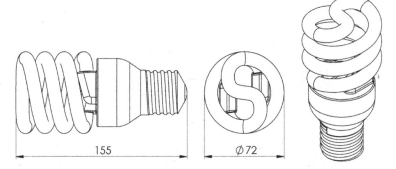

Fig. 7.28 Curve sweep on the model of an energy-saving lamp

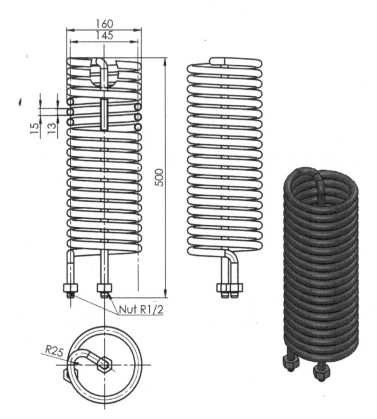

Fig. 7.29 Tubular heat exchanger

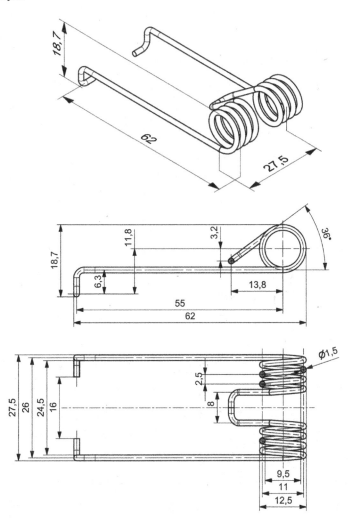

Fig. 7.30 Torsion spring

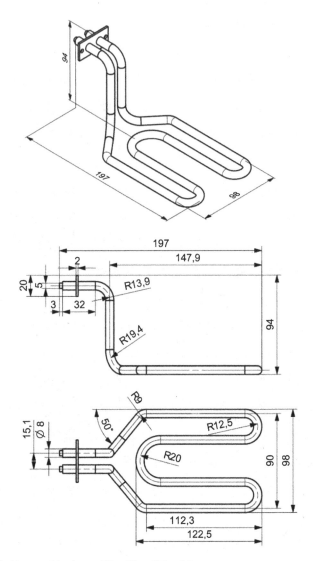

Fig. 7.31 Washing machine heater 70 × 98 × 197 × 94 mm

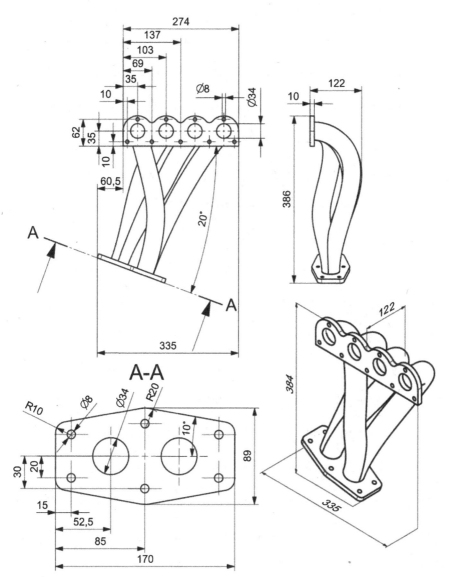

Fig. 7.32 Engine exhaust manifold 121 × 335 × 386 mm

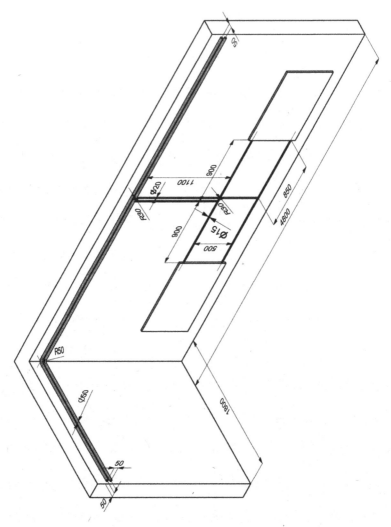

Fig. 7.33 Pipeline for central heating system with pipes $\phi50$, $\phi20$ and $\phi15$ mm

Chapter 8
Loft-Transition

Abstract Transitions over different cross-sections along curves are important for defining the shapes of spirals, i.e., hydraulic shapes in general. In this chapter, the shapes of spiral casings, typical of hydraulic machines, aerodynamic blowers, intake manifolds, low-pressure blowers, etc., are specifically defined. The examples show proofs of special influences, the poorly defined orientation of consecutive cross-sections, etc.

To create models that define a connection between two locations with different cross-sections, there is a special feature called loft. It is often used to design air-conditioning tubes, channels for directing airflow, liquids or free-flowing material. Using this feature it is possible to significantly shorten the time between designing, manufacturing and using a product. Furthermore, it allows the manufacture of different models with a changing shape of the cross-section.

8.1 Manufacturing Technology and Use

Transitions are created in continuous forms (castings and injected products) or they are segmented (individual manufacturing or products made of sheet metal, flat panels). Transitions can run in the direction of a vector or a circle. A common form of transition is a spiral casing, with a circle or s-shaped spiral as the guide direction, while the cross-section can be round, rectangular or square. If the manufacturing technology is casting or injecting, it yields a physical shape like that shown in Fig. 8.1a. However, with fewer items or just a single item, a segment technology is used. Segmenting is used for the transitions with sheet metal or plates from polymers, wood, glass etc. An example of spiral segmenting is shown in Fig. 8.1b.

Segmenting is also used for guiding free-flow materials with different transitions, such as from a square silo into a round flange with a closing lid, etc. The transitions along the lines of the transition are shaped by means of generating lines, connecting

DOI: 10.1007/978-3-319-03862-9_8, © Springer International Publishing Switzerland 2015

Fig. 8.1 Horizontal (**a**) and vertical (**b**) versions of the Francis turbine (*Source* http://en.wikipedia. org, http://www.tps.si)

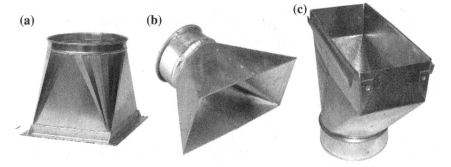

Fig. 8.2 Transition parts, made of sheet metal (*Source* http://www.nordfab.com). Loft square to circle (**a**), cylinder to rectangle (**b**), cylinder to rectangle prismatic form (**c**)

the points on different cross-section shapes, defined on the starting and ending planes (Fig. 8.2).

The key is to recognize the advantages of this feature. In our case, there is a digitized model which makes compiling the working documentation simpler. The blueprint includes information about the inlet and outlet cross-sections, defined by the inlet and outlet points, the inclination of the surfaces and the main dimensions. With such defined dimensions and the attached file of a digitised model, it is possible to start making a casting model or an injection tool using a CNC milling machine. A dimension check of such a created model or tool requires no further information. The same goes for the product itself.

However, if there is no digitized transition, the shape should include many more details.

8.2 Modelling Objects with a Variable Cross-Section

This modelling procedure can be seen as completing the curve sweep with a constant cross-section, as described in the previous chapter, but using a sweep where the cross-section is randomly, but continuously, transformed. In this case the transition between two or more cross-sections is performed by means of an interpolation between two shapes. As an example, let us take a square cross-section and a circle, not lying on the same plane, and create the transition between them. This transition can be described using a body in space.

8.2.1 Placing an Object in Space

Such 3D objects usually have a slightly more complex geometry, which makes the technological manufacturing also more complex. The geometry can be rather asymmetrical, coming close to an organic form. The method can be completed by the advanced modelling of the surfaces, where the emphasis is on the quality of the transitions between the cross-sections and the surfaces. The position of the object, created using this method, refers again to its natural position. The position of such an object in space is usually defined by logically placing the cross-sections in space before any transition or extrusion is performed (Fig. 8.3).

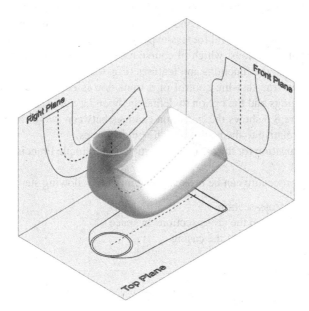

Fig. 8.3 Placing an object in space and the projections on the orthogonal planes

8.2.2 Creating a Base Sketch

As mentioned before, it is important in this modelling operation to choose the plane where the base sketches are placed. A transition is executed if the sketches of at least two transitions are not lying on the same plane. The sketching planes can be set at different distances, parallel or at an angle, but taking account of the geometrical constraints.

The basic parameters that are required for a transition with a variable cross-section are as follows:

- In the case of a guided sweep with a curve: one or more planar or spatial guides, continuous curves (spiral, B-splines)
- Setting one or more different cross-sections relative to the guide curve
- Adjustable transition parameters between sketches

8.2.3 Transition in Space

Similar to a curve sweep, transitions also allow open contours in the base sketches of individual cross-sections (Fig. 8.4).

According to the above-mentioned parameters that need to be included in the software, it is also necessary to take account of the method and the execution of the transition between the cross-sections. Some modellers provide the user with fewer options to set the transitions, and they execute an optimum transition in given conditions. This comes in useful for less-experienced users. Other modellers provide a vast number of transitions, which of course requires a higher level of experience and familiarity with the functions and features (Fig. 8.5).

It is definitely true that the control of a transition is better when the software allows more settings and the option to define the boundary guide curve, not just the centreline. Figure 8.6 shows the transition changes between two different elliptical cross-sections, relative to the position of the guide curves.

Below is a transition for the example of a model of a simple Francis spiral turbine casing (Fig. 8.7).

The modelling procedure can be broken down into the following steps:

1. Locating the cross-sections
2. Transitions between the cross-sections in space
3. Creating the inlet part of the casing
4. Creating the shell
5. Adding the central part
6. Adding the guide-vane blades
7. Completing the shape and the finished model

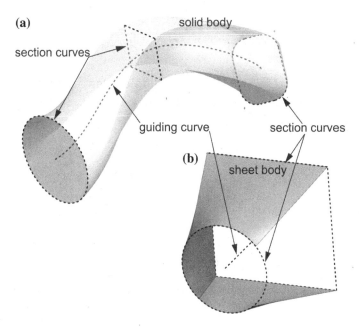

Fig. 8.4 Transition with a variable cross-section. Guiding curve (**a**), guiding line (**b**)

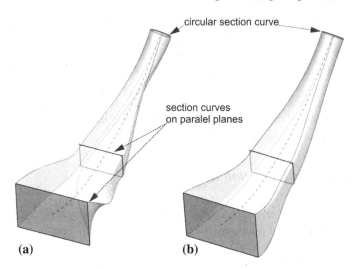

Fig. 8.5 Different transitions for identical cross-sections and an identical guide curve. Nonlinear cross-section oriented (**a**), linear cross-section oriented (**b**)

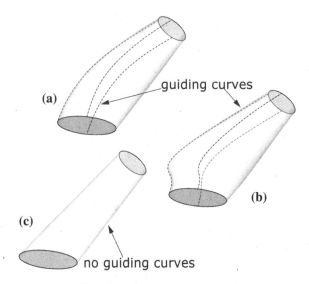

Fig. 8.6 Transitions at different boundary guide curves and without them. Simple guide vane curves (**a**), complex guide vane curves (**b**), linear or no guide vane curves (**c**)

8.3 Modelling in SolidWorks

8.3.1 Positioning the Cross-Sections

The procedure is similar to that of curve sweep. First, create the positions of the cross-sections in space, and then link them to obtain an object in space. It is important in SolidWorks to present each cross-section as an independent base sketch. Figure 8.8 shows the positions of the cross-sections when modelling the spiral part of the casing.

8.3.2 Forming a Transition in Space

A transition in space is performed using the *Loft* (*Insert > Boss/Base > Loft*) command. It is important that each cross-section is created as an independent sketch and then inserted into the set in a particular order. The basic form of executing transitions is presented in Fig. 8.9.

Practice has shown that using only a few cross-sections and a few sketches does not provide enough accuracy. For a more accurate description of a shape, a guide curve (*Guide Curve*) and/or a centre line (*Centerline Parameters*) must be added. For transition purposes, the centreline has a similar function as the path for the curve sweep function.

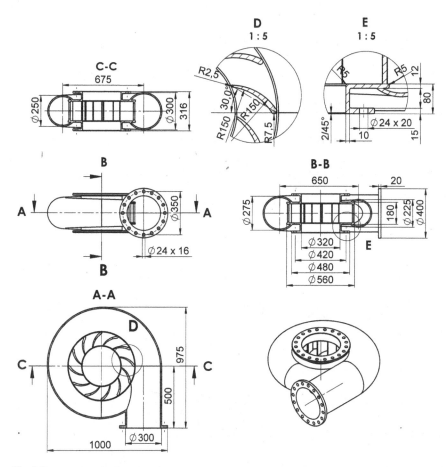

Fig. 8.7 A model of a Francis spiral turbine casing

In this case the centreline of the transition is created as a spiral curve (***Insert > curve > Helix/Spiral***). Set the spiral pitch (50 mm) and set the other parameters, as shown in Fig. 8.10.

Using the ***Loft*** command, execute another extrusion. Select the cross-sections that define their envelopes and apply them to form the transitions. Activate the centreline (***Centerline Parameters***) and select the prepared spiral (Fig. 8.11).

8.3.3 Creating the Inlet Part of the Casing

Create the inlet part of the casing. Use the default inlet cross-section (***Convert Entities***). Applying a linear extrusion, extrude it in a normal direction by 500 mm. Due to the tangential contact between the spiral part and the inlet tube, the software reports

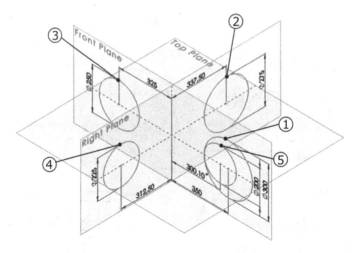

Fig. 8.8 Positions of the cross-sections for creating the spiral part of the casing. The *asterisk* next to the dimension represents an extra 0.1 mm due to the specific model

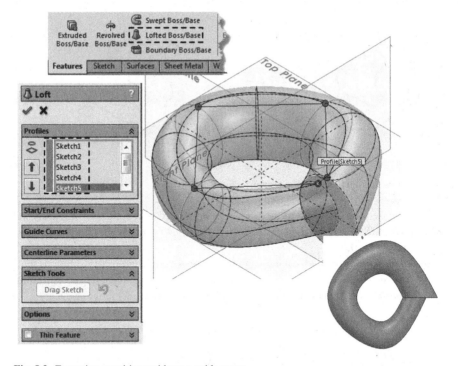

Fig. 8.9 Executing transitions without a guide curve

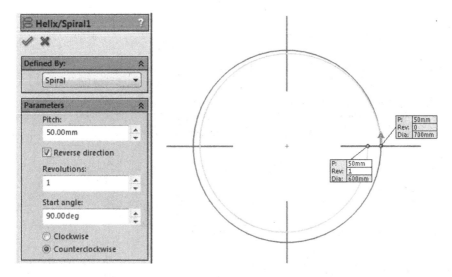

Fig. 8.10 The centreline of transitions, created by means of a spiral curve

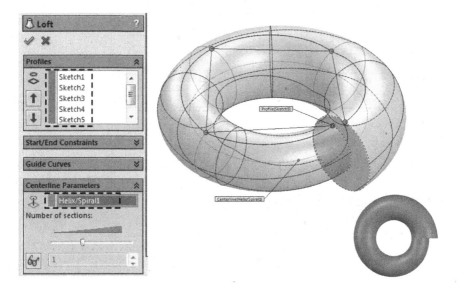

Fig. 8.11 Creating the transition by means of a centreline (*Centerline Parameters*)

a zero-thickness geometry problem (Fig. 8.12). To prevent this, make the cut first (*Extruded Cut*), and then add material (*Extruded Boss/Base*), using the same sketch (Fig. 8.13).

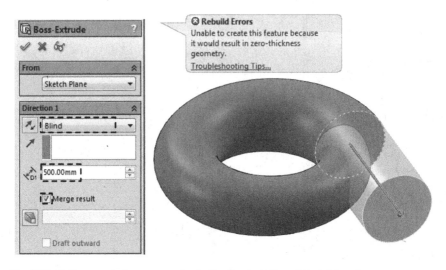

Fig. 8.12 Software reporting an error due to the direct addition of the inlet part

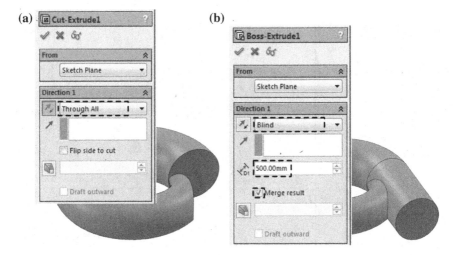

Fig. 8.13 Cutting (**a**) and adding (**b**) the inlet part

8.3.4 Creating the Shell

Create the shell (*Shell*) from an existing solid model (*Solid Model*) and use the (*Insert > Features > Shell*) command to create the shell elements. Set the shell thickness (10 mm) and mark the inlet surface that needs to be removed (Fig. 8.14).

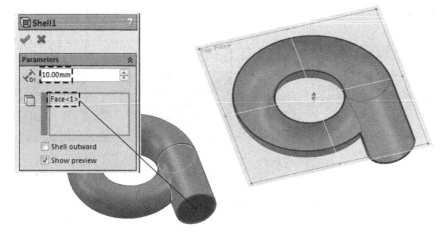

Fig. 8.14 Creating a thin-walled model and its cross-section

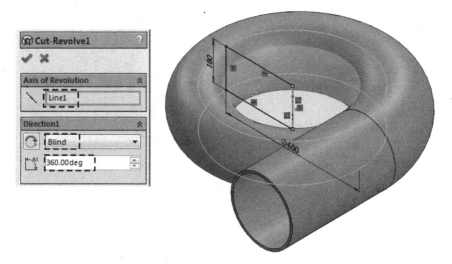

Fig. 8.15 Revolved cut for creating the central part of the casing

8.3.5 Adding the Central Part

In the central part of the casing, a guide must be created to direct the water flow. Using revolving, create the inlet cut out first (Fig. 8.15) and then we added central part (Fig. 8.16). When making the sketch, take account of the symmetry of the top and bottom parts.

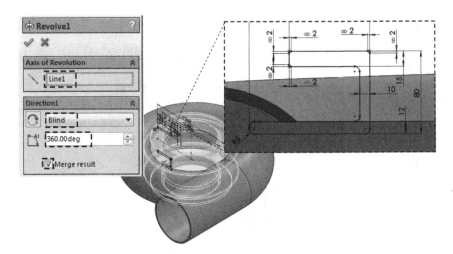

Fig. 8.16 Completing the central part of the casting with adding material

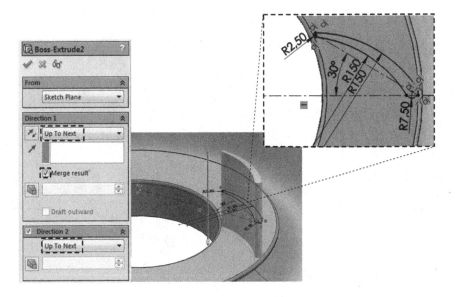

Fig. 8.17 Modelling a guide blade on a base plane

8.3.6 Adding the Guide-Vane Blades

In our case a simple guide vane with fixed blades will be modelled. The guide blades
are usually rotating. To model a guide blade, select a proper plane to draw a base
sketch—the top view of a ring in our case. The shape of the guide blade is defined
by the hydraulic calculation, taken in the form of the geometry in our case. Sketch

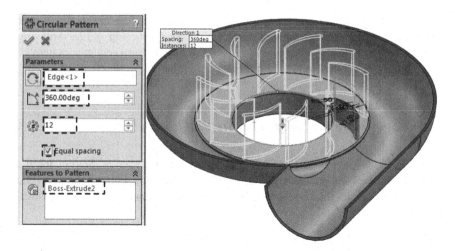

Fig. 8.18   Using a circular pattern for modelling the guide blades (n = 12)

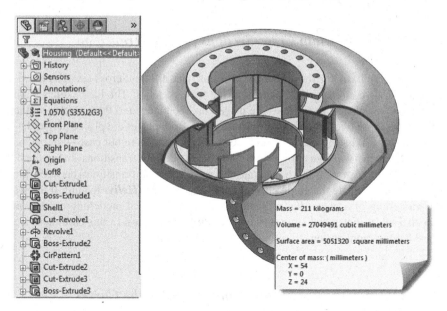

Fig. 8.19   A model of a Francis spiral turbine casing with the corresponding features structure and mass characteristics

the shape of the guide blade on the top plane, extending it by means of a linear extrusion up to the top and bottom surfaces (*Up To Next*) (Fig. 8.17). Applying a circular pattern (*Insert > Pattern/Mirror > Circular Pattern*) create a pattern of 12 repetitions along the entire circumference (Fig. 8.18).

8.3.7 Completing the Shape and the Final Model

Complete the casing model by adding bores with a diameter of 22 mm for the M20 bolt in the central part, and a flange to connect the tubing. Figure 8.19 shows the complete model with the corresponding features structure and mass characteristics. The material is steel S 355 (DIN 1.0570).

8.4 Modelling in NX

8.4.1 Setting the Main Cross-Sections for Loft-Transition

In our case the basis for forming the transitions with different cross-sections is represented by five circles, which represent the base sketches. The base sketches are placed on two planes: two circles are placed on one plane, and three circles on the other plane, with two circles, one inside the other. Both planes are placed perpendicularly to the z-axis (Fig. 8.20). Such base sketches—circles in our case—are placed in space in such a way that allows simply and clearly defined transitions among them.

For the purpose of a more accurate description of the transitions between the cross-sections, create the guide curve. In our case use the (*Helix*) feature. It allows the modelling of a planar helix (Sect. 7.4). For an identical modelling result you can also select the spatial splines (*Curve > Studio > Spline*). In this case there are considerably more options to reshape a curve in space.

8.4.2 Creating the Guide Curve and Orienting the Curves of the Cross-Sections

The path of the guide curve can be controlled with two parameters: (1) the degree of the spline and (2) the tangent condition for the curve at the intersections of the cross-sections. Qualitatively, you can make use of the curvature indication on the curve, which can be defined with the (*Analysis > Curve Analysis*) command (Fig. 8.21). In this case, follow the curve before and after the selected point by means of the (*Studio Spline > Type > Through points*) command. The entire curve is defined by consecutively defined points, generally defined by their position on

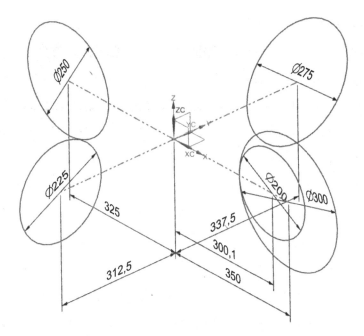

Fig. 8.20 Dimensions and the positions of the contours of the starting cross-sections, ready for extrusion

the cross-sections, specified in the base sketches. For the simultaneous completion of the points assembly, you can choose a position on the curve, and then—with a double click on the mouse (*MB1*)—decide whether to add or delete a selected new point. This can contribute significantly to the shape of the transition between one and another base sketch. The (*Studio Spline*) command offers the possibility to specify two constraints (*Constraints*), already mentioned above. For each new or existing point, we can define the continuity (tangent) relative to any axis of any coordinate system. In our case we will guide the curve in the control points so that the tangents at these points will be perpendicular to the cross-sections (Fig. 8.22).

Using the (*Swept*) feature, you can model the transitions between cross-sections and use them to define the inner part of the transition. For a more accurate specification of the transitions, other parameters also need to be strengthened. First, define the sections where the transitions will be formed (*Sections*) (Fig. 8.23). By selecting the cross-sections one after another in this menu and clicking (*MB1*) on their contour, identify the contour and then confirm it with a click (*MB3*). After each confirmation you can move onto the next identification of a new cross-section. By doing this, you make a list (*List*) of cross-sections that will define a 3D body once the transitions between them are carried out. The next parameters make it possible to define the guide curve—one or a maximum of three. First, take a look at the created curve, created by the software according to the conditions of the cross-sections, set in space, and running through the centres of the circles on individual planes. In our case, this is the

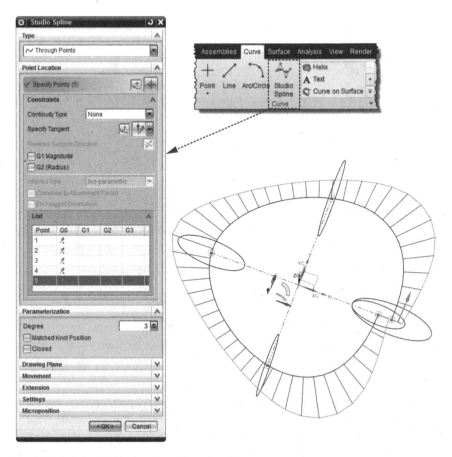

Fig. 8.21 Defining the guide curve with cubic splines

spiral around the z-axis. In this step, you can verify the orientation and the starting points of the circles or curves, which in effect defines the connection between the local coordinate systems and the global one. This is to verify the parallelism and connections of the generating lines on the transition surface (Fig. 8.24).

For a quick preview, make use of the (*Swept > Preview/Show Results*) command that makes it possible to display the created 3D object. When forming transitions, represented by their outside surface, the incorrect orientation of the curves results in a twisted or completely deformed shape. Pinches appear when orienting points leap by 180°, and minor defects of the surface can occur if the orientation is, for example, staggered by a few or a few tens of degrees (Fig. 8.25). Correct the starting point of the circle or the curve by activating the edit menu of the base sketch (*Edit/Sketch*) and then identify each circle with the cursor and confirm it with a double click (**MB1**). Then find the menu (*Arc/Circle/Full Circle*), tick it and change the beginning and the end of the curve (Fig. 8.26).

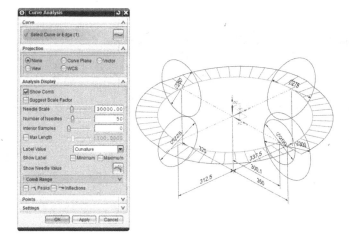

Fig. 8.22 Analysing the curve that defines the transitions between the individual cross-sections by displaying the curve radii with the (*Curve Analysis*) command

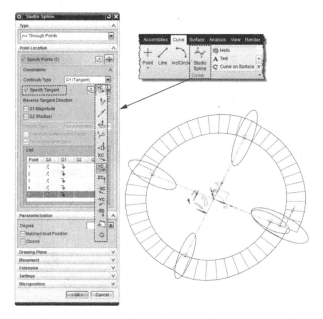

Fig. 8.23 Defined guide points of the curve where the tangent is specified (G1)

After the control, perform the operation of forming consecutive transitions (*Feature > Swept*) by selecting once again all of the previously mentioned required parameters, and creating transitions along the guide curve (Fig. 8.27).

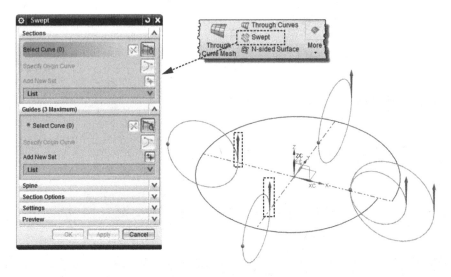

Fig. 8.24 A look at the orientations of the individual cross-sections on the base planes from one to another consecutively placed cross-sections prior to the last step of defining a transition

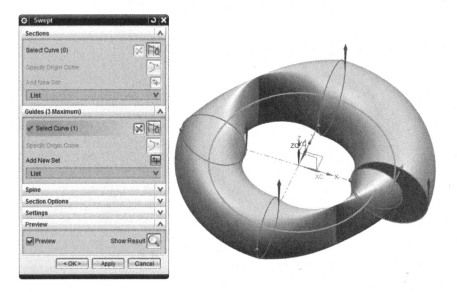

Fig. 8.25 Execution of a transition in the case of staggered starting points for individual cross-sections. The staggered cross-sections result from the incorrect orientations of individual circles or—generally speaking—curves that define cross-sections

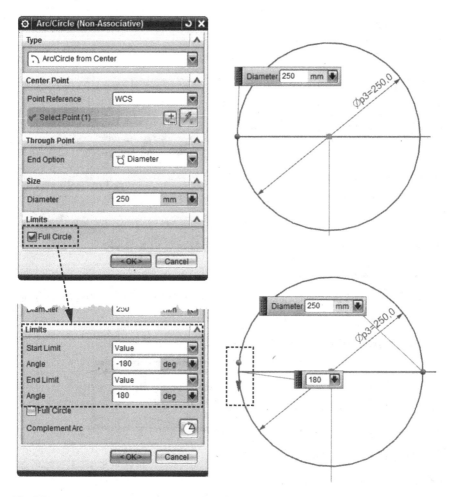

Fig. 8.26 Changing the starting point of a curve or a circle in the base sketch

8.4.3 Creating a Transition Through Cross-Sections by Means of the Guide Curve

8.4.4 Completing the Shape

The next phase involves making a shell from a solid model. This is carried out with the (***Feature > Shell***) command. This feature requires specifying the area for removal. Material is added to the rest of the surface model by projecting the surface, running parallel for a certain distance. The direction of the normal relative to the surface also needs to be given. On our model, direct the normal inwards and select a thickness of 10 mm (Fig. 8.29).

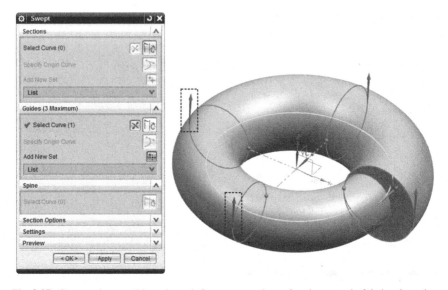

Fig. 8.27 Consecutive transitions through five cross-sections after the control of their orientation

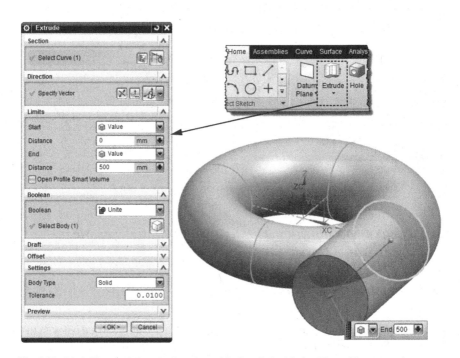

Fig. 8.28 Modelling the inlet tube into the turbine's spiral with the (*Extrude*) command

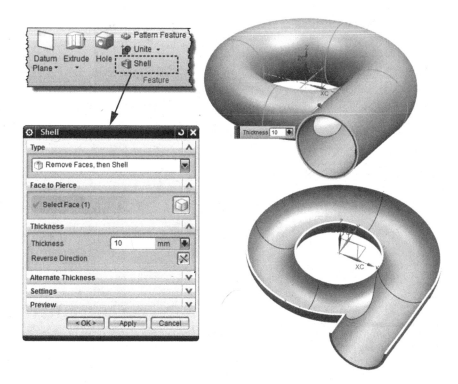

Fig. 8.29 Modelling the shell with the (*Shell*) command

The spiral casing is then modelled by cutting out the centre part by means of a rotational rectangle of 325 × 180 mm, drawn on the XC-ZC plane.

When modelling the opening in the centre of the spiral casing, a flange should be modelled. Modelling the casing's flange begins with a new base sketch, positioned on the front plane of the XC-ZC spiral (Fig. 8.30). The base sketch should therefore be modelled in that plane in order to be able to execute the revolving around the z-axis of the main coordinate system (*CS*). In the sketch itself, the profile is mirrored along the z-axis of a two-dimensional coordinate system (*Sketch/Mirror Curve*) (Fig. 8.31).

Attach the flange to the spiral casings with the (*Boolean/Unite*) function.

Equip the 3D object with details, created with the patterning functions. In the next step the sketch of a pre-guide blade will be created (Fig. 8.32) underneath the casing and extruded with the (*Extrude*) feature.

Having created the basic pre-guide blade by means of circular patterning, proceed by creating all twelve blades. For this modelling use the *Feature > Pattern Feature/Circular Layout* command. Set the number of blades to 12 and the full angle of the area of even distribution, which is 360° in our case (Fig. 8.33).

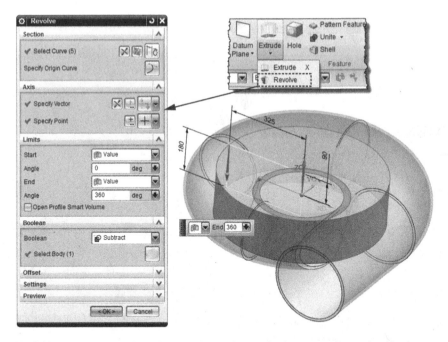

Fig. 8.30 Removing material in the central part of the casing by means of a rotational body

In its rough form, a model should have all the characteristics of the shapes, spec-ified by the basic function. The connecting elements also belong to the functional specification of a product. In our case they are represented by the bolt connections on the flanges. This requires modelling holes on the casing's flange. This is done with the (*Feature > Hole*) feature. When positioning the holes, their location should be precisely defined (*Specify Point*). When processing this command, a sketch will open where the position of the centre of the bore (hole) should be specified. The selected surface, lying in the positive direction of the XC axis, allows the centre of the hole to be defined, located 210 mm from the centre of the flange (Fig. 8.34). The hole has a diameter of 22 mm for an M 20 screw. The modelling of the bores should be finished completely, i.e., all the bores on the flange. This is performed using the (*Pattern feature*) feature. When the holes on the horizontal flange above and under-neath are defined, there are 20 holes. The holes on the inlet flange are modelled in a similar way. Because the inlet flange has a smaller nominal diameter, there are only 16 holes of the same diameter (Fig. 8.35).

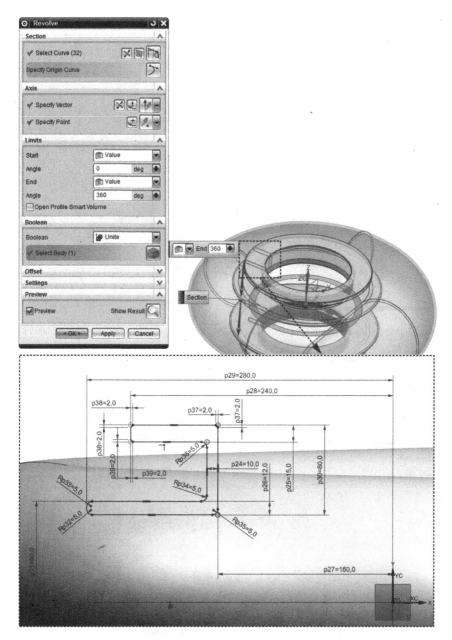

Fig. 8.31 Flange detail in the base sketch, enabling—by means of a rotating body (revolve)—a 3D model of the spiral casing's flange

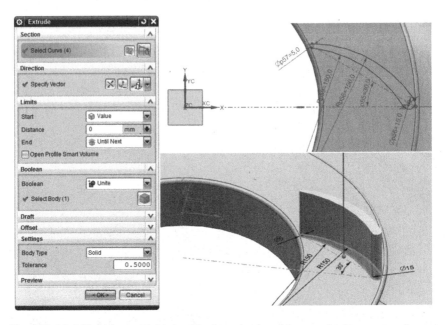

Fig. 8.32 Modelling a pre-guide blade with a base sketch and the extrusion between two flanges

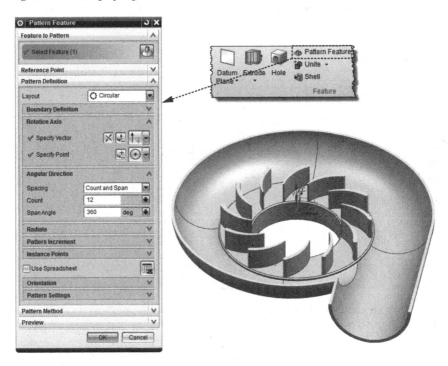

Fig. 8.33 An example of a circular pattern for a pre-guide blade with a feature (*Feature Pattern*)

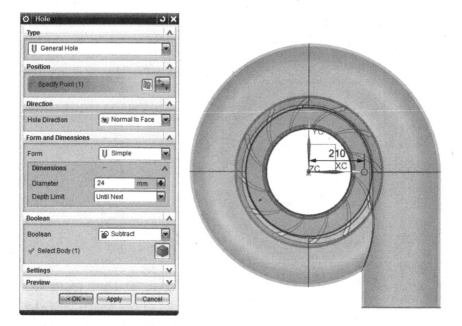

Fig. 8.34 The position of the point to create a hole of 22 mm in diameter for an M 20 screw with the (*Hole*) feature

As a rule, the nominal diameter is the inner diameter of the tube. It is also referred to as the hydraulic radius and is important for calculating the tensions and the flow capacities.

Final model is presented on the Fig. 8.36.

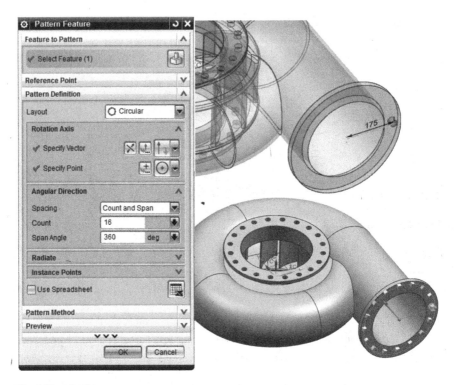

Fig. 8.35 Inlet flange

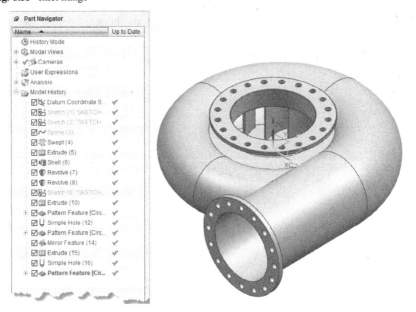

Fig. 8.36 Final model of Francis spiral turbine casing

8.5 Examples

Figures 8.37, 8.38, 8.39 and 8.40 show few examples of lofted parts.

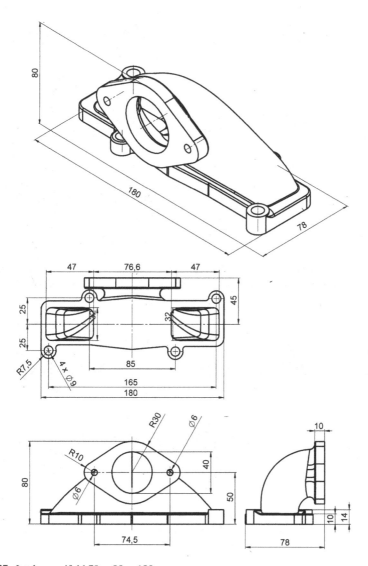

Fig. 8.37 Intake manifold 78 × 80 × 180 mm

Fig. 8.38 The external shape
of a simple handset

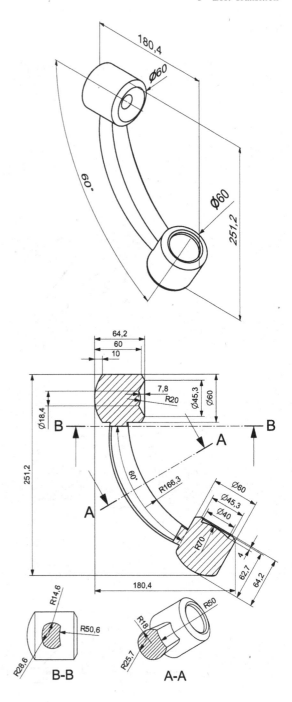

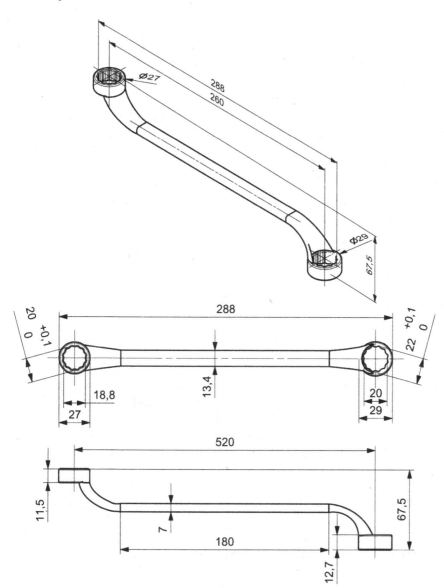

Fig. 8.39 A ring wrench

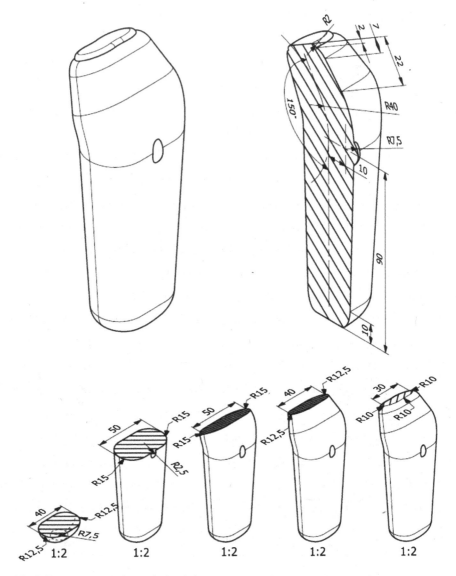

Fig. 8.40 The external shape of a hand shaver

Chapter 9
Supplementing the Shape

Abstract Special detailed shapes, defined by chamfers, edge blending and extrudable surfaces are a key part of any detailed design. For all of the shapes dealt with in the previous chapters, the use of supplementary shapes on different models and technologies is described. This chapter includes the specifics of the presented features, important for top-level design, and in particular for creating photographs of non-materialized products. So far we have discussed the modelling of relatively rough product shapes. These shapes have not included the specific details that are used in practise and that consider the technological and detailed shapes in terms of their structure. They also express the perfection of the product, which is referred to as excellence. Below we will present the modelling of detailed shapes, usually resulting from the use of certain technologies. This type of modelling has different names in different environments: from combined shape, to complex shape, to designing for an X technology. In fact, a model in all its details should be created by following a real product and a real technological process.

This chapter will present the features that result from a particular modification and can therefore not be used independently. These features need an existing basic geometry of a model that is then technologically redesigned, which is why the term supplemented (applied) shapes is used.

This chapter will also show the use of applied features with an emphasis on the modelling of castings and injected parts.

9.1 Manufacturing Technology and Use

The supplemented shapes' features are intended for the detailed design of models - products for casting or the injection of plastics. For a better understanding, the first presented item is a complex casting (Fig. 9.1) with a technical drawing, showing drafts for extracting the casting from the sand.

J. Duhovnik et al., *Space Modeling with SolidWorks and NX*,   221
DOI: 10.1007/978-3-319-03862-9_9, © Springer International Publishing Switzerland 2015

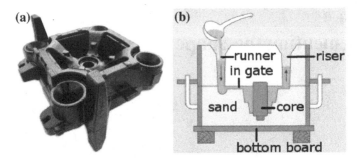

Fig. 9.1 A semi-complex casting with drafts towards the dividing plane (**a**), the principle of sand casting with defined draft planes (**b**)

Fig. 9.2 Plastic box for waste material: tool for injection molding (**a**) and final product (**b**)

With polymer injection, drafts are considered to a lesser degree because of the material with a smaller elastic modulus E (N/mm^2), which makes it more deformable. Furthermore, it is also possible to make taller products with very small drafts. For lower heights, a product with larger drafts is easier to extract from the tool (Fig. 9.2).

Internal pressure is used for the special technological procedures of injecting products from plastic materials. It allows the high-quality formation of a wall thickness without an inner core (Fig. 9.3).

9.2 Auxiliary Shapes in the Modelling Process

Supplementing the base shape means supplementing the base shape of a product in terms of a selected or a specific technology, used for its manufacturing. Each product should be shape-adjusted using certain functions in such a way that it makes the

Fig. 9.3 Injection moulding with internal pressure (small draft)

manufacturing possible. This step is important for the execution and specifies the details. Also available to the user is a set of features and commands that support the geometry and upgrading of the existing features in such a way that details are defined according to the chosen technology. Below is a list of the general features that are of some importance and are used for each completion.

9.2.1 Fillet

Blending the external edges improves the safety of a product in use; it also looks smaller and not so rough. When modelling fillets, new surfaces (one or more) should be added to the marked models. The operation of geometric assembly is performed by keeping the geometric body consistently and completely closed with both the surfaces and the basic volume.

When several consecutive edges are marked the programmes allow automatic adjustments of the other edges in accordance with the settings. The latest modellers include several other advanced, fillet-feature settings (Fig. 9.4). Besides the basic setting of a constant radius along the entire edge, more advanced programmes also allow the setting of a variable radius along the edge (Fig. 9.5a) that can also have a conical finish with a different radius or a different radius having the value 0 (Fig. 9.5b). In the corners of a 3D object's edges it is also possible to set a corner setback. This defines the transition between the edge blends at the intersection of three faces (Fig. 9.5c). Modellers usually include a default setting to obtain a tangential joint between the straight and rounded parts of a face. Edge-blend modelling is used in upgrades in order to obtain pronounced transitions between the surfaces and to change the degree of the derivations.

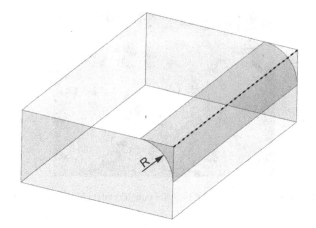

Fig. 9.4 Applying a fillet (*Fillet/Blend*) to a selected edge with a pre-set radius

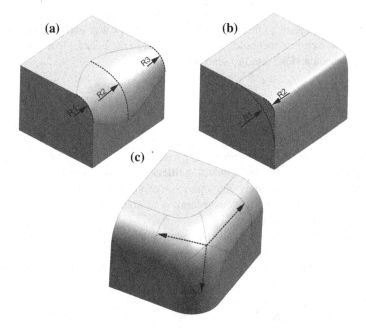

Fig. 9.5 A few examples of the additional possibilities of modelling a larger number of edges with all the finished edge blends

9.2.2 Chamfer

The easiest geometry for redesigning edges is the chamfer. For chamfer modelling the procedure is similar to a fillet, with the only difference being that the edge is placed on the side flat surface (Fig. 9.6).

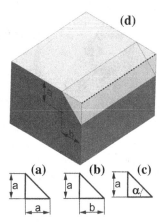

Fig. 9.6 Submitting the starting parameters to create a chamfer: a = const (**a**), $a \neq b$ (**b**), a and α (**c**)

For a chamfer, the following parameters should be submitted (Fig. 9.6):

- Chamfer depth *a* when the chamfer is symmetrical
- Depths *a* and *b* when the chamfer is not symmetrical
- Depth *a* and chamfer angle α

9.2.3 Shelling

Shell (**Shell**) is an important feature for specific technologies that are nowadays used for injecting plastic materials and very complex castings. This feature makes it possible to carve material out of a 3D solid body or to open it for access from one or several sides through a selected surface. This is performed by erasing selected surfaces. After erasing, the remaining surfaces automatically acquire the pre-set wall thickness, either inside or towards the outside. The surfaces become thicker, depending on either the positive or negative direction of the normal on the surface (Fig. 9.7).

Logical and geometrically defined constraints apply to this feature. The offset (**Offset**) itself cannot be repeated indefinitely. The constraints are shown in Fig. 9.8. In the first part they are dependent of the details' constraints. The limited radii lengths should be taken into account, especially when approaching the value 0, and on the other hand, when the values exceed the length of the edges where these radii are integrated into the original, base shape. Problems occur wherever the offset exceeds the values that are specified by the described limits and then the software will report an error. The thicken (**Thicken**) feature is similar to that of shelling. Thickening is, in principle, the opposite feature, allowing us to model a solid body by means of offsetting (**offset**) a rectangle on the surface.

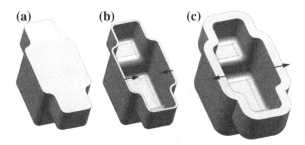

Fig. 9.7 Using the shell feature and defining the direction of thickness with the normal to the selected surfaces: full solid model (**a**), thickness vector to the mass center (**b**) and thickness vector on the opposite way of the mass center (**c**)

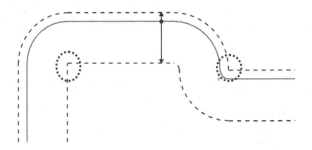

Fig. 9.8 A scheme of possible surface problems due to radii constraints

9.2.4 Ribs

Ribs—often imposed by a requirement for mechanical reinforcement or a faster filling of cavities in the injection mould—are modelled by means of extrusion with some conditions. The most common approach is to model a rib with a base sketch, giving the rib's base shape, or a line, specifying its position or section. The basic rib shape can be extruded on a sketch plane or perpendicularly to this plane (Fig. 9.9).

9.2.5 Draft

A draft is of vital importance to determine the method of extracting a product from its mould (metal, sand or polymer). The draft (**Draft**) feature is so designed that it is first necessary to specify the draft of a selected surface relative to a stationary face. The draft is graphically specified by the orientation of a vector, and numerically by a draft angle (Fig. 9.10).

The draft feature is important because it can also be applied to the most complex shapes, taking account of the constraints of previous features and other geometric constraints (Fig. 9.11).

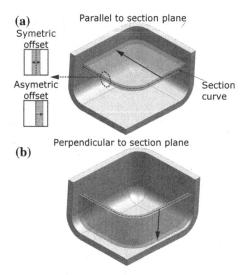

Fig. 9.9 An example of using a rib modelling feature: rib parallel to the section plane (**a**) and rib perpendicular to the section plane (**b**)

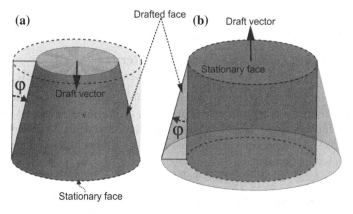

Fig. 9.10 Creating the draft (*Draft*) and the required parameters: stationary face from *bottom up* (**a**) and stationary face from *top down* (**b**)

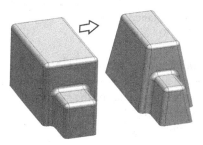

Fig. 9.11 Creating the draft of a complex geometry in one step

Fig. 9.12 An example of
the linear patterning (*Linear
Pattern*) of the bores on a
panel (face): principle of
linear pattering (**a**) and modell
of linear pattering (**b**)

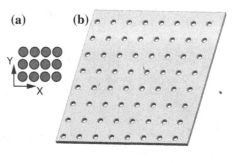

Fig. 9.13 An example of
a circular pattern (*Circular
Pattern*) for a convex pattern
on a round plate: principle
of circural pattering (**a**) and
model of circular pattering (**b**)

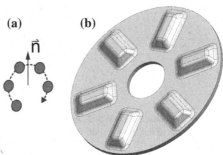

9.2.6 Patterning Geometric Entities (Pattern)

Patterning is important because it makes it possible to draw a shape (bore, hole, element, etc.) only once and then transfer it once or several times further on to pre-set faces according to a certain order or pattern. The shapes are defined by the geometric entities (surfaces, bodies, curves, etc.), and replicating them requires defining their attributes before patterning.
The basic examples of patterning are well known:

- **Linear pattern** (Fig. 9.12)

  Patterning is possible along one or more axes on a given face. The parameters of importance are: (1) the offset between the patterns (2) the direction or number of patterns on the prescribed distance.
- **Circular pattern** (Fig. 9.13)

  A circular pattern is defined by a vector with a given orientation, followed by rotating a selected pattern around it. The most common parameter for defining a new position of the pattern is the angle and the number of the pattern or the number of patterns along the entire circumference.
- **Curve driven pattern** (Fig. 9.14)

Fig. 9.14 Patterning along
the guide curve: principle of
curve pattering (**a**) and model
of curve pattering (**b**)

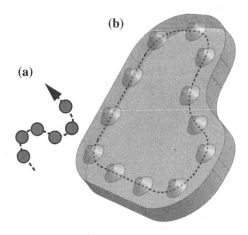

The same approach can be applied to patterning along a selected, pre-set curve. In this case, select a curve, followed by distributing the pattern along the curve, along separation units, or by distances.

Modern programmes include larger sets of patterns. Patterns can be features, geometric entities, such as bodies, surfaces etc. As shown later, a similar approach can also be adopted at the assembly level.

9.2.7 Mirroring Geometric Entities (Mirror)

The mirroring feature allows geometric shapes and features to be mirrored about a selected axis (on a plane) or plane (in space) (Fig. 9.15).

An example of a simple plastic cover will be presented (Fig. 9.16).
The modelling procedure can be broken down into the following steps:

1. Creating the base shape
2. Adding the reinforcing ribs
3. Creating the holes
4. Creating the drafts
5. Adding the fillets

9.3 Modelling in SolidWorks

All modellers tend to include supplemented shapes in such a way that it makes coming to the end result as easy as possible. You should of course follow the highlighted, automated procedures that accompany the technological features and have an impact

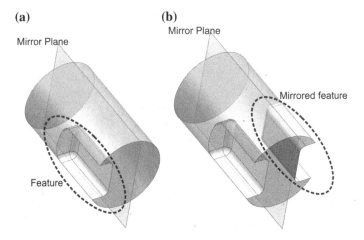

Fig. 9.15 Mirroring geometric shapes about a plane: feature mirroring (**a**) and mirrored feature (**b**)

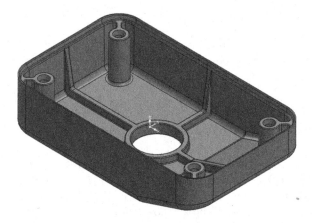

Fig. 9.16 An example of a simple plastic cover

on the whole set of edges, faces or volumes that make presenting an object possible. The auxiliary shapes in SolidWorks are user-friendly and easily included when redesigning an object. First, create a base shape and supplement it to create the end shape by means of applied features (***Applied Features***).

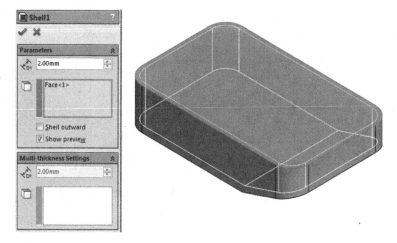

Fig. 9.17 Modelling the shell element, representing a cover with the main dimensions

9.3.1 A Base Model

The base shape of a cover comes from the function that it is intended for. The space to be covered is represented by a volume in the shape of a rectangular solid with the dimensions $100 \times 60 \times 24$ mm, lying on the top plane. The vertical edges are rounded (***Fillet***) with R = 10 mm. The cover is shown as inside out for easier modelling of the inside. In our case, rotate it by 180° and on its back. Having finished the modelling, put it back in its natural position. This is one of those examples where a model (product) is not in its natural position but rotated around the axis so that we have the clearest view of the object. On the front part (the bottom in our case), apply a chamfer (***Chamfer***) to the edge for a length of 30 mm and a height of 10 mm. The shell (***Shell***) is created with a thickness of 2 mm with the top face removed (Fig. 9.17).

On the front part of the cover (the bottom in our case) there is a hole with a diameter of 20 mm, thickened on the inside with a ring that has a thickness of 2 mm and a height of 3 mm (Fig. 9.19). Such a design detail is characteristic of the reinforcements on covers with large flat faces. Attach the cover onto the surface by guiding it with the edge, where the cover thickness decreases to 1 mm. To fix it, use a connection depth of 2 mm. To serve this purpose in our case, a positioning edge with a thickness of 1 mm and a depth of 2 mm is created on the top. On the inside part of the cover, four screw sockets with a diameter of 8 mm are created. They allow screwing with fixing screws, aligned on the top part with the bottom face of the fold. The position of the sockets, i.e., their centres in the upper plane, are the same as the centres for the curve radii of the outside vertical edges. Create one socket first and replicate it three more times in the upper plane, using the (***Linear Pattern***) function (Fig. 9.18).

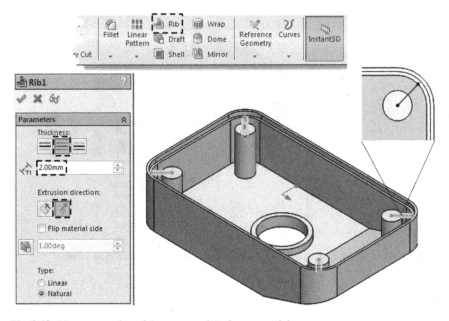

Fig. 9.18 Linear patterning of the screw sockets for a screw joint

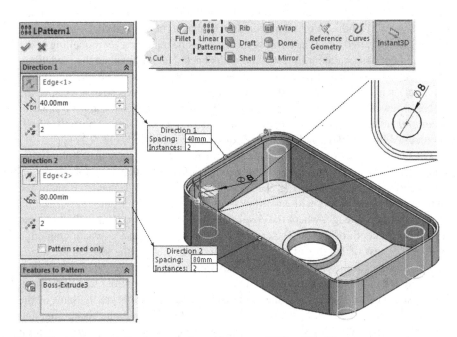

Fig. 9.19 A base sketch for the rough shape of a cover of 100 × 60 × 24 mm

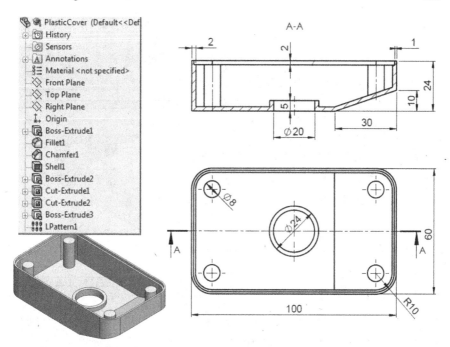

Fig. 9.20 Adding reinforcing ribs between sockets and the wall

9.3.2 Reinforcing Ribs

Reinforcing ribs serve several functions (**Rib**): (1) the cover's stiffness is increased in its basic rectangular shape, which is important for fitting, and (2) the die-casting abilities of the sockets improve as it is easier to die cast them over the ribs, which reduces the die-casting and injection times. Figure 9.20 shows the adding of reinforcing ribs between the sockets and the edge blends. Due to the principle of an identical wall thickness, the rib thickness is identical to the thickness of the cover's basic wall (2 mm).

On relatively large, flat surfaces, reinforcing ribs are often added on the inside. In our case, the ribs will be placed in the shape of a cross in order to improve the stiffness on the inside of the cover. Reinforcing ribs are modelled progressively, beginning at the back (Fig. 9.21) and then at the front (Fig. 9.22). Transverse ribs are modelled last (Fig. 9.23). The modelling procedure shows that with the model being symmetrical, relative to the longitudinal (front) plane, it is possible to apply the mirror function (Fig. 9.23).

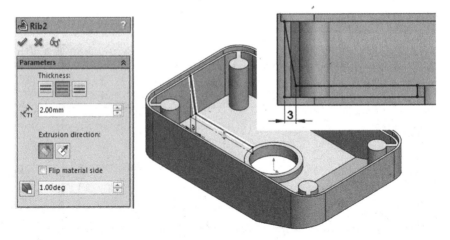

Fig. 9.21 Longitudinal reinforcing rib at the back

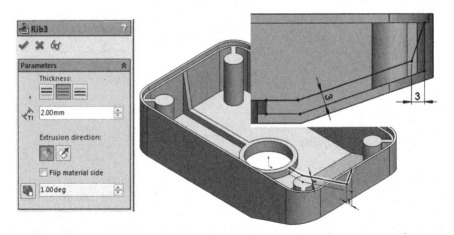

Fig. 9.22 Longitudinal reinforcing rib at the front

9.3.3 Attachment Holes

The sockets must be added holes that allow the screwing of self-tapping screws. For this purpose, SolidWorks has the hole wizard (*Hole Wizard*) in its menu, allowing the rapid and easy modelling of different types of holes and bores. You need to select a type of bore, specify its parameters, and position them on the model (Fig. 9.24). These user primitives are an important addition to specific technologies and the user can integrate them into the modeller software.

Due to the specific manufacturing technology, select a countersunk (*Counter-sunk*) hole and specify the parameters (Diameter: 4 mm, depth: 2 mm distance from the bottom face, angle: 30° and chamfer diameter: 5 mm) (Fig. 9.24).

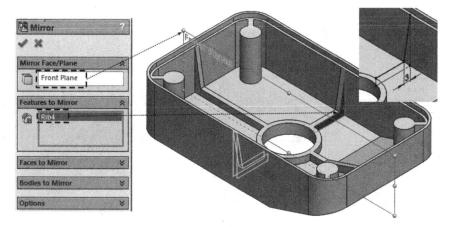

Fig. 9.23 A transverse reinforcing rib and mirroring about the longitudinal bisectional plane

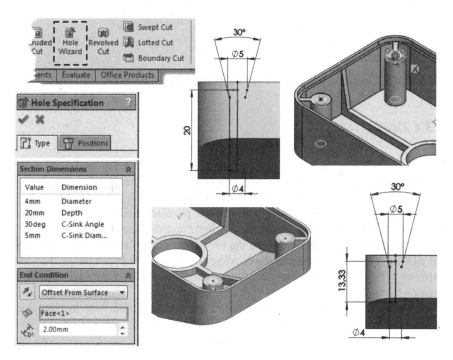

Fig. 9.24 Using the wizard to create holes in the screw sockets for the attachment with self-tapping screws

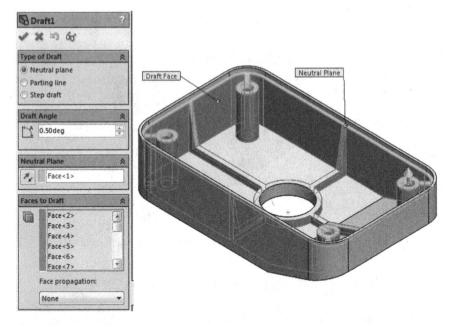

Fig. 9.25 Draft modelling in the direction of the dividing plane on the model

9.3.4 Draft

Because a product needs to be extracted from the tool, there are different drafts to be used, enabling the extraction in the direction of free movement. Drafts are modelled with the (*Draft*) feature. Define a suitable draft and the dividing plane of the product in the tool or forms (die casting). With the dividing plane selected, specify the surfaces you wish to draft. In our case, select the dividing plane as the contact plane between the cover's face and the frame, i.e., on the top of the sockets, and create a 0.5° draft for all the vertical surfaces under the dividing plane in order to open the product for extraction (Fig. 9.25). For easier coupling of the cover to the base frame, execute a draft of 10° at the guide edge.

Because it is possible to forget to define the drafts for all the faces due to their large number, all modellers include the function of surface draft control or controlling the product's extraction from the tool. In SolidWorks the control is performed with the (*Draft Analysis*) command (Fig. 9.26).

9.3.5 Fillet

It has been pointed out several times that detailed chamfering and edge blending are specified only for a completely formed shape. This applies also to a draft. Details

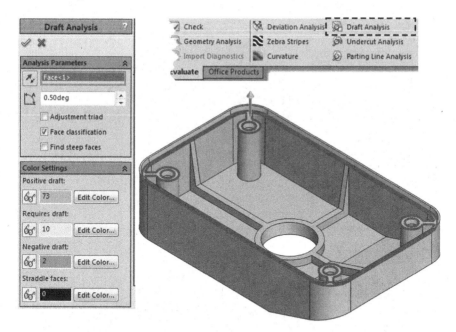

Fig. 9.26 Draft control in the direction of the dividing plane on the model

can be rounded only when a model has been dimensionally complete and the drafts performed. Rounding the base shape can be performed earlier. This rounding (edge blending) is performed with the (***Fillet***) command. For different rounding radii, one set of values will be grouped together and the other set also. First, the outside part will be rounded (R = 4 mm) (Fig. 9.27), followed by rounding on the inside part (R = 2 mm) (Fig. 9.28). Add the edge blends on the sockets (R = 1 mm) (Fig. 9.29) and finish by rounding all the passages with a fillet of 0.3 mm (Fig. 9.30).

As a special addition, let us not forget finishing the holes for the self-tapping screws in the sockets, which is performed in a spherical shape as the hole is executed with plugs (note that this is about die casting or injecting, not drilling or milling). The plug crowning is executed with a command for special shapes (***Dome***) (Fig. 9.31).

9.3.6 Final Model

Figure 9.32 shows the complete model of a polymer cover with the corresponding structure of features and the mass characteristics.

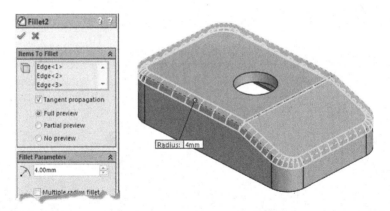

Fig. 9.27 Rounding the bottom part R = 4 mm

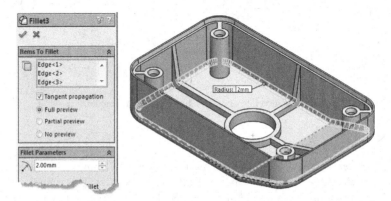

Fig. 9.28 Rounding the inside edge R = 2 mm

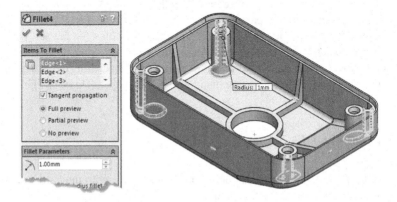

Fig. 9.29 Rounding at the fixing plugs R = 1 mm

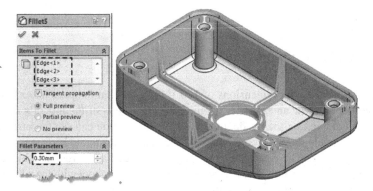

Fig. 9.30 Rounding the remaining edges R = 0.3 mm

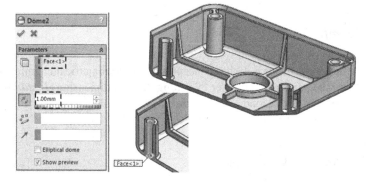

Fig. 9.31 Modelling crowned surfaces at the end of the holes in the sockets

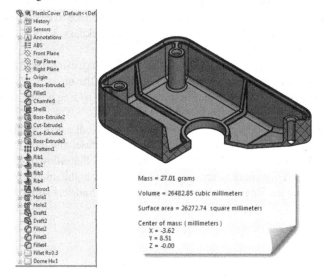

Fig. 9.32 A model of a polymer cover with the corresponding structure of features and the mass characteristics

9.4 Modelling in NX

9.4.1 Creating a Rough Model

Define a base cube with dimensions of $100 \times 60 \times 24$ mm in space in such a way that it will later allow an easy view of the individual parts of the model's shape. If a model is complex on the inside, then model it so that you can see the inside, and when the outside's shape is more complex, the main view should be focused on the outside (Fig. 9.33). The base shape has very pronounced and large edge blends, so these should be completed as early as possible on the base model with the edge-blend (*Edge Blend*) feature (Fig. 9.33, bottom).

9.4.2 Chamfer

Due to the requirement for a design concept with an asymmetrical chamfer, begin with applying the adjustable chamfer by setting the first dimension (*Distance 1*) at 10 mm, followed by the second one (*Distance 2*) with a value of 30 mm (Fig. 9.34).

9.4.3 Carving the Cover (Shell)

Carve the model by using the (*Feature > Shell*) feature, which results in a cover of adequate thickness. The procedure begins by defining the face to be carved, followed by defining the direction (thickness vector) and the thickness. In our case, select a thickness of 2 mm (Fig. 9.35).

All the details will later be subject to the familiar extrusion (*Extrude*). At the centre of the front plane there should be an opening in the cover. Create the round opening by removing a cylinder and in the next step by adding a ring, modelled by means of the extrude function (Fig. 9.36).

Specify on the model the plane up to which the guide edge is deepened, followed by specifying the edge thickness, which is 1 mm in our case, and then move it down by 2 mm (Fig. 9.37a). Fix the cover, using the sockets that allow the screwing of self-tapping screws. Cylindrical screw sockets are placed in the middle of the outside rounding of the cover. Begin by drawing a pattern of cylindrical screw sockets with $\phi 8$ (Fig. 9.37b).

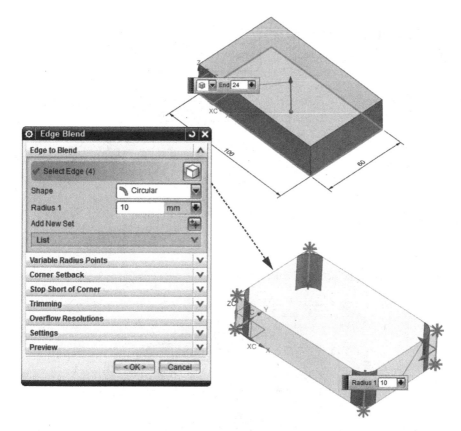

Fig. 9.33 Modelling the base shape with a large edge blend R = 10 mm (*Edge Blend*)

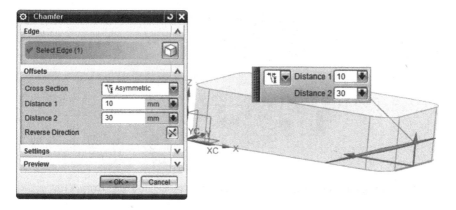

Fig. 9.34 Supplement the model by chamfering (*Chamfer*) on the bottom edge (the model is inside out in order to expose the complex shape on the inside)

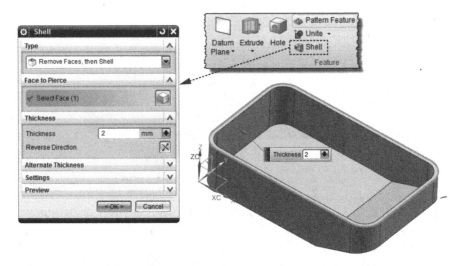

Fig. 9.35 A model after shelling with the *(Shell)* feature

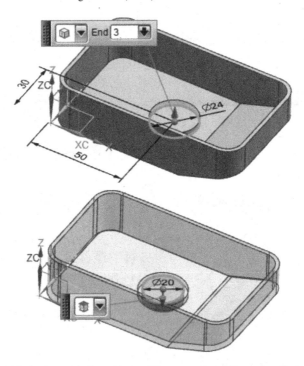

Fig. 9.36 Modelling a through hole and hub reinforcement by adding a ring (*Extrude*)

Fig. 9.37 Modelling the cover's guide edge and the pattern of the cover's attachment plug: cover's guide edge (**a**) and cover's attachement (**b**)

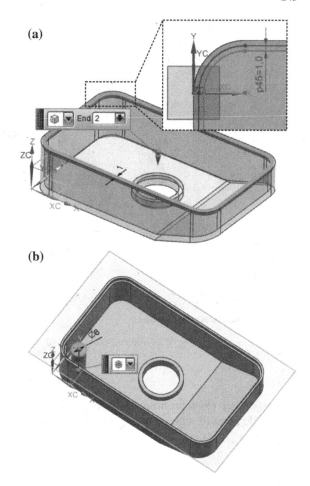

9.4.4 Patterning Geometric Entities (Pattern)

The modelling of a pattern, a cylindrical screw socket in our case, is performed in two directions (**Pattern Feature > Linear**). In the first direction (**Direction 1**) set the pitch distance at 80 mm and the number of the pattern repetitions at 2. For the other direction (**Direction 2**), set the pattern pitch distance at 40 mm and the number of repetitions again at 2 (Fig. 9.38). The number of repetitions means the number of all the repetitions after the pattern operation has been executed.

There are also other ways of patterning, accessible with the following commands:

- **Pattern Geometry**—This command allows the patterning of finished geometric entities (curves, surfaces, bodies, etc.)

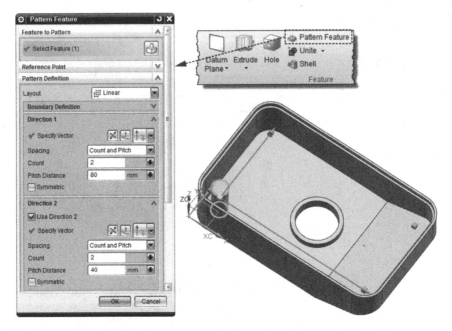

Fig. 9.38 Linear patterning of a geometric entity in two directions (*Pattern Feature*)

- *Pattern Face (Synchronous Modelling)*—This command allows the patterning of single faces or a group of faces (this command uses less memory, making its execution faster)

9.4.5 Rib Modelling (Rib)

When modelling with this feature, create a base sketch first, lying on a perpendicular plane relative to the rib extrusion (Fig. 9.39). Where and when to begin modelling the first rib is up to the user, so make a plan of the modelling procedure. In our case, begin by modelling the rib between the outside wall and the fixing screw socket. The base sketch on the perpendicular plane relative to the rib extrusion can be executed with the (*Perpendicular to Section Plane*) command.

The ribs reinforce the front plane and are placed at the centre between the outside walls and connect the hub of the round opening and the outside wall. In this case, the (*Parallel to Section Plane*) command is the most useful one. It specifies the longitudinal rib shape (changing height)—the rib shape along its length from the beginning to the end of the reinforcement. With the extrusion specified, and no rib thickness, the latter should be specified with the (*Rib > Walls/ Parallel to Section Plane > Symmetric/ Thickness*) command. In our case, the thickness is 2 mm. The

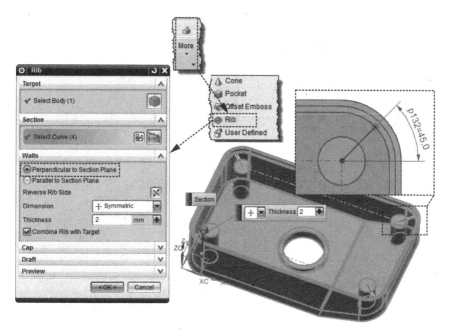

Fig. 9.39 Modelling connection ribs (**Rib**) between the fixing screw sockets and the outer wall

rib sketch is shown in Fig. 9.40. When modelling reinforcing ribs, make use of the geometry projection (bottom thickness) by using the command in the rib sketch specification menu (**Project Curve**). When using an auxiliary geometry, there can be a problem with the execution of the extrusion, which requires changing the auxiliary geometry into the reference geometry (Fig. 9.41).

9.4.6 Mirroring About a Plane (Mirror)

Each feature that is symmetrical about the middle plane of any model can be mirrored with the (**Mirror Feature**) command (Fig. 9.42). Mirroring is often used because in nature there are usually mirrored images in at least two axes. This set of commands also allows several options, depending on the geometric entities being used:

- **Mirror Geometry**—This command allows the mirroring of finished geometric entities (curve, surfaces, bodies, etc.)
- **Mirror Face (Synchronous Modelling)**—This command allows the mirroring of one or more faces (this command uses less memory, making the execution quicker)

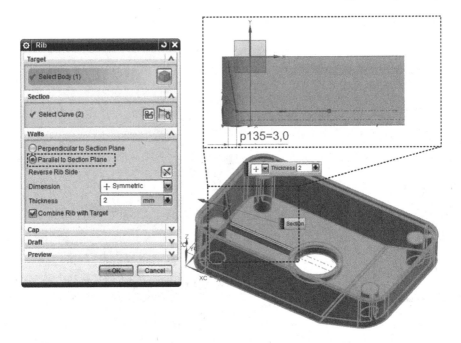

Fig. 9.40 Modelling reinforcing ribs (**Rib**) between the hub and the outside, and the cover's top outside wall

9.4.7 Creating Bores (Bore–Hole)

The (**Feature > Hole**) feature is specifically intended for creating bores (including standard ones). In contrast to the general cylindrical extrude (**Extrude**), this feature allows the creation of various standard thread bores by simply putting a 2D sketch point into the position where a bore should be (Fig. 9.43). In our case, self-tapping screws will be used, which requires the use of a bore shape that allows a continuous load transfer from a metal screw into a polymer screw socket.

As the fixing is executed with self-tapping screws, select a countersunk (**Counter-sunk**). In the menu, select the order of the dimensions as shown in Fig. 9.43. Make the holes in the sockets on the cover's upper part, followed by the same procedure on the lower part (Fig. 9.44).

9.4.8 Draft Modelling (Draft)

As mentioned previously, the cover will be made out of a polymer, using injection-moulding technology. This technology demands drafts in the direction of the injection tool's dividing plane, which means that the drafts should be formed up or down from

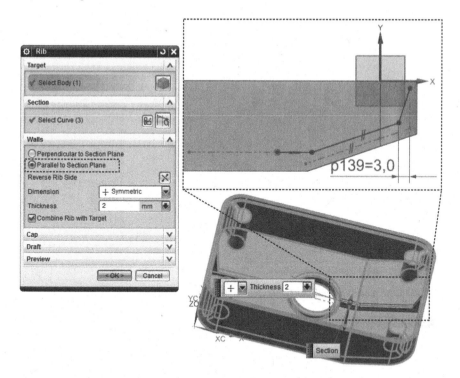

Fig. 9.41 Modelling reinforcing ribs (*Rib*) between the hub and the cover's bottom outside wall

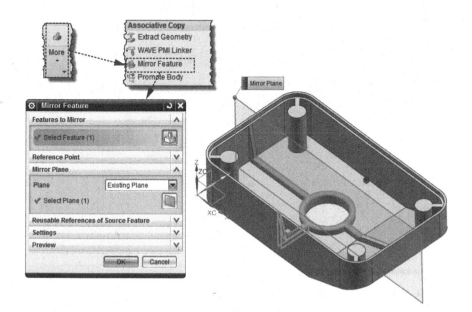

Fig. 9.42 Modelling reinforcing ribs from the hub to the side of the cover's outside wall

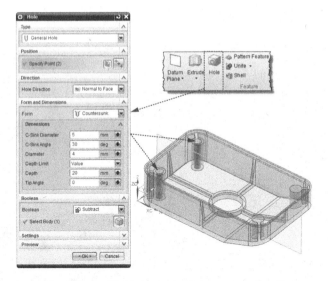

Fig. 9.43 Bore modelling with the (*Hole*) feature for fixing screw sockets

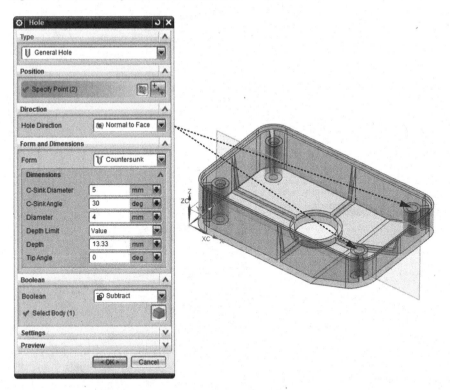

Fig. 9.44 Bore modelling on the lower screw sockets

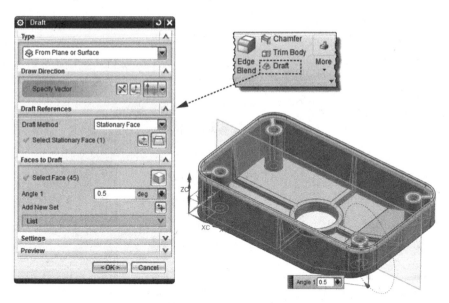

Fig. 9.45 Modelling a draft of 0.5° (*Draft*) for all the vertical surfaces, lying under the dividing plane

the selected dividing plane, depending on the direction of the model. The model can be directed towards either the bottom part or the top part, or towards both parts. If a model is not tall it can be directed in one direction only, towards one part of the tool. If it is tall, use the drafts in both parts of the tool. It reduces the changed dimensions of the product by half.

In order to define the dividing plane, specify the draft vector. The vector is perpendicular to the dividing plane. It is referred to differently, depending on the modeller, and is specified with the (*Stationary Face*) command. In our case, the model is low, so the guide edge's plane is also used as the dividing plane (Fig. 9.45). The dividing plane, selected in the (*Draft References*) menu, is referred to as the stationary plane. It is logical that it is not necessary to perform drafts on all the surfaces, and not all the drafts are generally the same, which is why individual model surfaces can be selected with (*Select Face*). The selected surfaces are then given draft angles. There is rule that the draft is smaller for the outside dimensions and larger for the inside ones because injected objects shrink, increasing the forces required for the extraction of the product from the tool (Fig. 9.45). This is referred to as a positive draft. The correctness of the draft direction is checked by testing the negative edges and surfaces. For low heights, drafts are executed at a wider angle. In our case, this recommendation is used for the guide edge draft, where a draft of 10° is used for the height of 2 mm (Fig. 9.46).

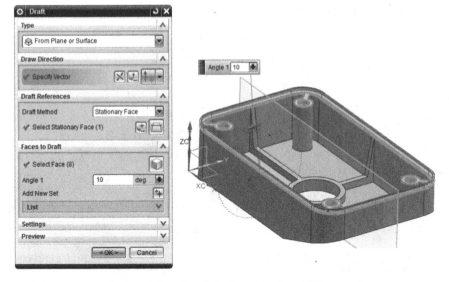

Fig. 9.46 Additional draft modelling (*Draft*) for the guide edge of 10°

9.4.9 Adding Detailed Edge Blends

It was mentioned before that modelling functionally defined edge blends comes at the beginning of the modelling process. All the edge blends that provide functionality and are related to a particular technology are defined at the very end because detailed edge blends are relatively smaller than the changed geometry of the model's faces or parts. It is for the designer to decide how to approach defining the detailed edge blends. Our advice is to model the edge blends from the highest values to the lowest ones. To do that, they have to be specified separately for each model. In our case, we will opt to first do the edge blends on the cover's front walls. The ergonomics and the industrial designer's conditions are met by performing edge blends with a radius R = 4 mm (Fig. 9.47).

In the above-described edge-blend procedures, different values for different types of edge blends with different conditions were used for specifying their size. So, it makes sense to classify them into groups in advance and prepare a special tree structure (*Part Navigator*) (Figs. 9.48 and 9.49). This is executed by right-clicking (*MB2*) the menu to choose a feature group (*Feature Group*). For reasons of simplicity, groups are named with clear marks, for example: Blend R = 0.3 mm (Fig. 9.50).

At the end of the modelling, make sure to check whether the model is complete in all its details, which is why it is sensible to use the feature structures and for drafts, checking the negative edges.

In our case it is clear that the model is fairly complete; it is only missing the bore ending in the sockets. It is flat, which could result in thermal stresses on the round edge on the plugs in the tool. It is these plugs that make sure that the bore

Fig. 9.47 Adding edge blends (***Edge Blend***) R = 4 mm on the cover front

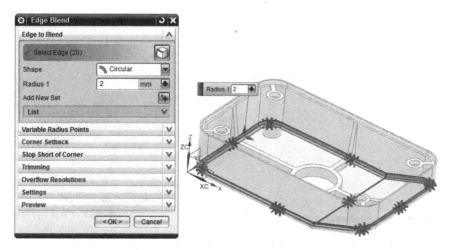

Fig. 9.48 Adding edge blends R = 2 mm on the cover's inside edge

for the self-tapping screws is prepared in advance. It is technologically sensible to relieve the thermal stress on the plug, to make a semi-circular plug ending and thus significantly relieve the thermal stresses on the sharp edges and reduce the extraction forces. Round the plug with the (***Edge Blend***) command to obtain an ending in the form of a hemisphere. The easiest way to do this is by rounding it with a diameter of the same size as the plug. In order to be precise, measure the tip of the plug first. In the radius-specifying menu (***Radius 1***) use a right click (***MB2***) to activate the menu and select the (***Measure***) command. In the next step, measure the circle radius at the plug's end (note that plugs of greater lengths are conical; 1: 20, 50). Programme-

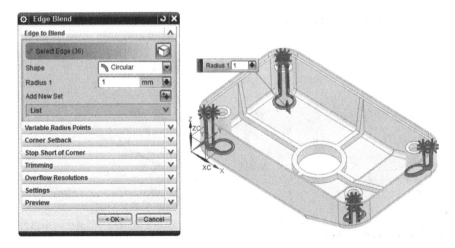

Fig. 9.49 Adding edge blends R = 1 mm at the root of the screw sockets and connecting the ribs between the screw socket and the cover's outside wall

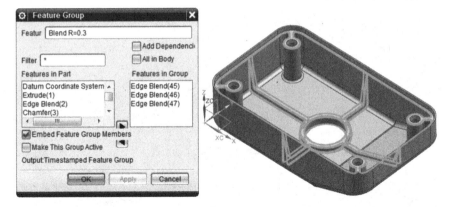

Fig. 9.50 Adding the remaining edge blends by forming a group. Edge blends are created due to the technologicality of the injection process and extracting the product from the tool

wise, the procedure is so defined that the measured value becomes the default entry value for the radius (Fig. 9.51).

Having finished the cover modelling, check the structure of the features and their order (Fig. 9.52).

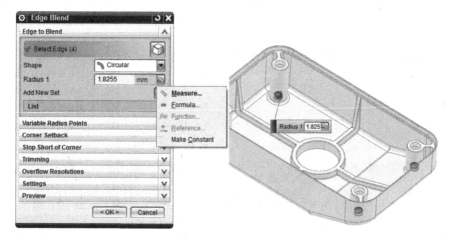

Fig. 9.51 Measuring the radius at the end of the plug and the procedure for the default radius value

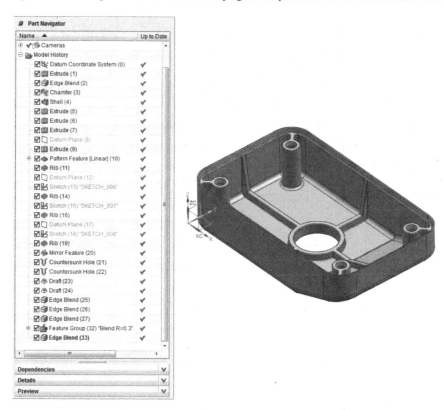

Fig. 9.52 Final model of a polymer cover with the features' tree structure

9.5 Examples

Figures 9.53, 9.54, 9.55, 9.56 and 9.57 present few examples of supplementing the shape..

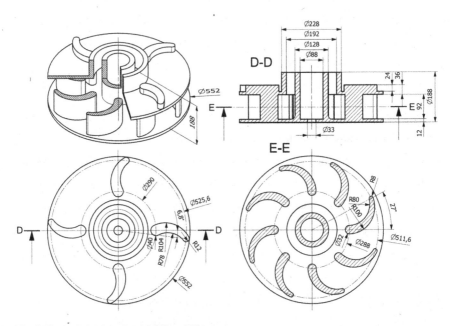

Fig. 9.53   Radial drive wheel ϕ552 × 188 mm

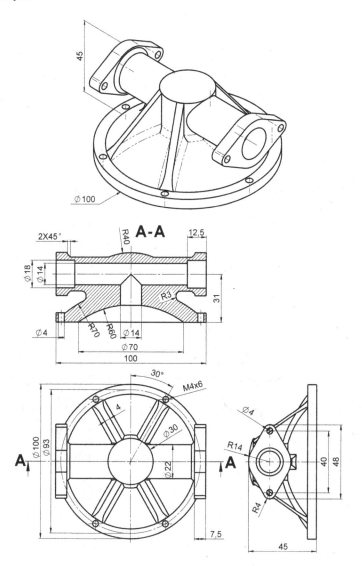

Fig. 9.54 Pressure vessel $\phi 100 \times 45$ mm

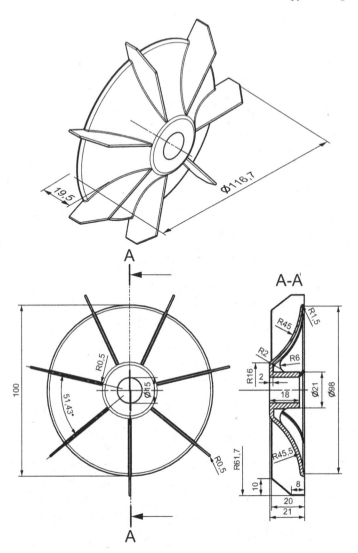

Fig. 9.55 Axial-radial rotor $\phi 120 \times 21$ mm

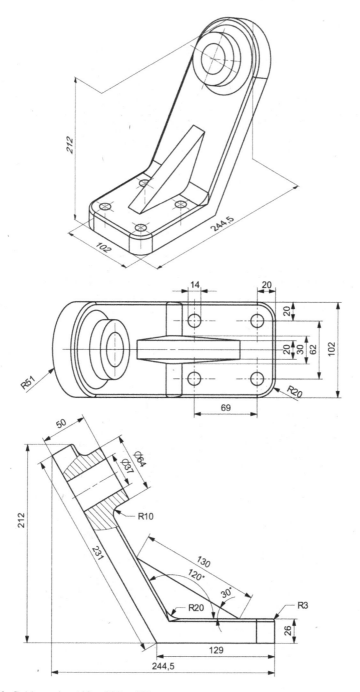

Fig. 9.56 Guide casting 102 × 220 × 235 mm

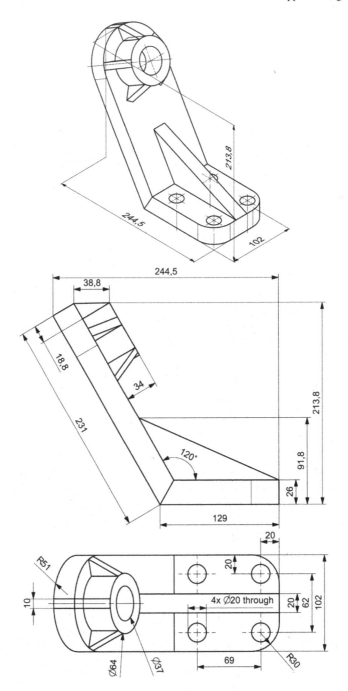

Fig. 9.57 A welded guide 102 × 220 × 235 mm

Chapter 10
Welding a Construction

Abstract Welded products are a special design case. While the features in the previous chapters were dedicated to rolling, die casting and the injection of plastic materials, welding is important for load-bearing structures. This is the reason why the features that provide a realistic presentation of welded details that define the volume of the material to be welded, the length of welds, etc., are important. Two concepts are presented in detail, i.e., the automated (pre-set) and specific (detailed, special) design of the welds. This provides the reader with the possibility of using both concepts on a single steel structure and thus modelling a real environment or shape.

Welding technology demands specific shapes that are not only determined by the welded joints, but also by the use of intermediate products (sheet metal, profile, cast, forged part, plastic injection, etc.). A welded assembly as a product can have two different shapes in the structure of the technical documentation. The old version of the technical documentation informational treats a welded assembly as an individual part. The more recent informational scheme considers it as a system of individual parts that are elaborated in the manufacturing drawing.

In our case, we include the welded assembly in the common drawing, but the individual parts are defined in the manufacturing drawing. Our common drawing is formed on a *Multi-body part* principle. It is essential to first define the individual parts that are formed by the additional processing of standard profiles (pipes, square pipes, full hexagon, etc.) in the drawing and later weld them into the final shape by composing. We mark the potential additional processing on the common drawing so that it is clear what is produced before and what is produced after the welding.

Our aim is to present the modelling of welded constructions from standard profiles.

J. Duhovnik et al., *Space Modeling with SolidWorks and NX*,   259
DOI: 10.1007/978-3-319-03862-9_10, © Springer International Publishing Switzerland 2015

10.1 Manufacturing Technology and Use

Welding processes and joining in general differ regarding the width of the joint of the used sheet metal. Similar processes are used with the width of the sheet metal: (1) thin sheet metal with a width between 0.3 and 3.0 mm, (2) medium-thick sheet metal with a width between 2.5 and 16 mm, (3) thick sheet metal with a width between 12 and 50 mm and (4) thicker sheet metal with a width between 40 and 200 mm. We weld a joint with similar sheet metal in a direct contact or with at least three times the width. The thinnest sheet metal in the joint should not be thinner than 0.3 times the width of the thickest sheet metal. Possible exceptions are static loads or pulsar loads, but never alternate loads. The same also applies to the width of the welded assembly "a", that is $0.3 > a > d$, where "d" is the width of the thickest sheet metal of the welded joint.

We have presented these few validities because it is important that the welded construction is modelled logically and with regard to all the conditions of the techno logics. The welded construction should not be drawn in any case, even if someone has good drawing skills or if someone likes to draw.

We can perform the welding with various technological processes and devices. Subsequently, we present the most common examples. Anybody that is going to work on welded constructions should deepen their knowledge, because the welding process itself brings a specific energy to the structure. This is also the reason why welded constructions generally have residual stresses that sometimes reach the level of the breaking stresses.

Welding is a technologically difficult operation in terms of positioning the electrode and has an impact on the environment and on the person doing the welding (Figs. 10.1 and 10.2). For this reason we also use robots, which provide a total digitalization of the model (Figs. 10.3 and 10.4). Robots can help the welder as well as creating high-quality and reproducible welds.

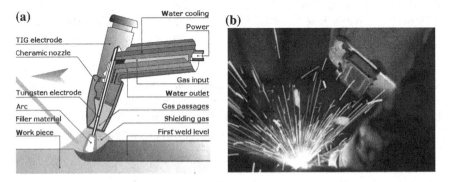

Fig. 10.1 Principle of the TIG process: TIG electrode in details (**a**) and view on the TIG process (**b**)

Fig. 10.2 Performance of a butt weld

Fig. 10.3 Detail of a welded joint

10.2 Modelling Welded Constructions

10.2.1 Modelling Welded-Beam Constructions

In the modelling of constructions from welded trusses or beams it is typical that the trusses or beams are usually the profiles of various standard shapes. The most common way of modelling these kinds of objects is building the system from a spine that we connect with the preferred standard or differently defined profiles. In static constructions in mechanics we first define a static model of the steel construction. In the modelling we compile the static model as a skeleton model of the steel construction.

In principal, the spine is a set of touching lines that we form with 2D or 3D sketches (Fig. 10.5).

The user can define standard profiles on his or her own (Fig. 10.6) or they can be imported from libraries of standard parts.

In welded constructions we need to carefully control the force transitions or the material tensions in the joints. In mechanics we refer to these as nodes. We always

Fig. 10.4 Welding the corner weld on a cover using two robots that have adaptive steering for the position of the electrodes

(a)   **(b)**   **(c)**
welded beam construction   static model   beam construction skeleton

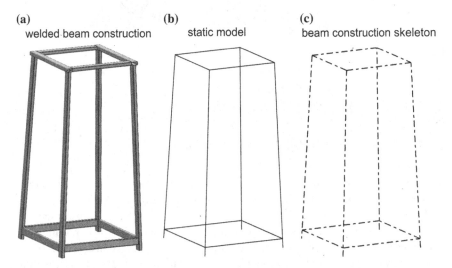

Fig. 10.5 Model of the welded construction of *rectangular hollow* profiles 40 × 20 × 2 mm (**a**), a static model of the welded construction (**b**) and a skeleton structure (**c**)

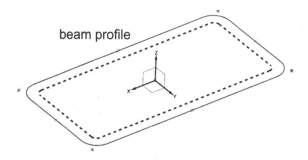

Fig. 10.6 Definition of the profile in a sketch (*Sketch*)

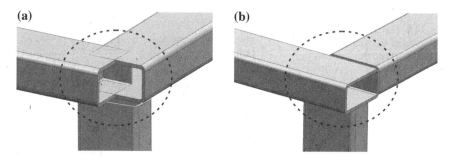

Fig. 10.7 Modelling the shapes of profiles in junctions of the spine in such a way that force transitions are provided through the material connections: middle ended profile (skeleton dimensions) (**a**) and construction ended profile (stress closing detail) (**b**)

need to form transitions separately because it is difficult to ensure the connections of force transitions by designing the model since they are defined in mechanics. This kind of automatic shape of the node transformations is only possible with tension and deformation control and at the same time the use of the laws of mechanics. It is not possible to form nodes in this way by using modellers that only consider the geometry. We usually arrange the nodes and the connections of various profiles in the last phase of the modelling of welded constructions (Fig. 10.7).

We can define the connections of profiles in the joints by standard shapes or we can perform them uniquely if we have newly derived details. Most modellers have a special module for this kind of modelling. In this case, standard operations like extrusion and revolving are used, which are customized by special features that enable rapid process. In addition to beam constructions we can use pre-set modules for modelling various electric installations or pipe systems.

It is important to emphasize that there is a difference between the individual systems. In welded constructions we model in one file (*part*) with multiple closed bodies (*multi-body parts*); however, for example in the case of a pipeline system it is already an assembly. This will be presented in subsequent chapters.

Fig. 10.8 When we produce
welded constructions we do
not combine individual 3D
bodies into a whole

Welded multi-body part

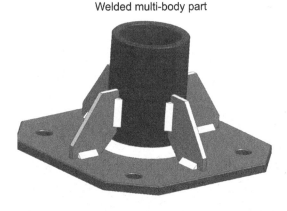

Fig. 10.9 Difference between
the model of the welded construction (**a**) and the common
3D model (**b**)

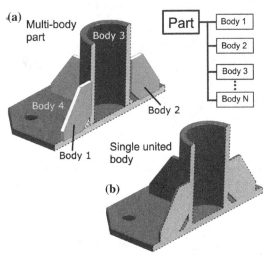

10.2.2 Modelling Other Welded Constructions and Marking Welded Assemblies

In the system of one file, which is defined for one part, welded constructions form closed bodies (multi-body parts) (Fig. 10.8). Later in a 2D sketch this enables us to separate the individual parts of the welded assembly and display them individually in a sketch of the welded assembly.

Figure 10.9 shows the difference between the common modelling (Fig. 10.9a), where we combine individual 3D objects and features in a joint body, and the modelling of the welded assembly, where we leave individual bodies separated (Fig. 10.9b).

(a)

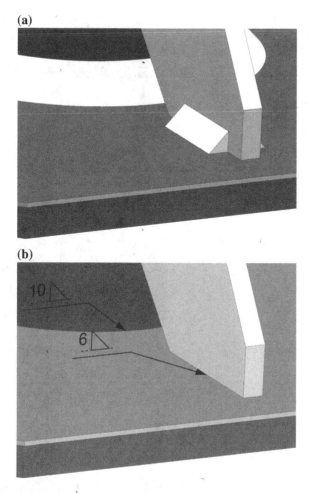

(b)

Fig. 10.10 The spatially modelled welded assembly (**a**) and the signified welded assembly (**b**) that considers the characteristic of designing the construction due to welding processes

When we want to highlight the shape of the welded assembly we model it as an individual part of the volume. We can also signify welded assemblies on the object in 3D space (Fig. 10.10). For this purpose many software packages have special user models.

In order to present the use of modelling steel constructions we use the chair in Fig. 10.11. Later, we will only consider the steel construction of the chair as shown in Fig. 10.12.

We can summarize the process of modelling the steel construction below:

1. The sketch of a static model of the steel construction—sketching a skeleton
2. The shapes of the profiles
3. The extrusion in space

Fig. 10.11 Produced massive chair that has a frame similar to a steel construction

Fig. 10.12 Steel construction of a chair

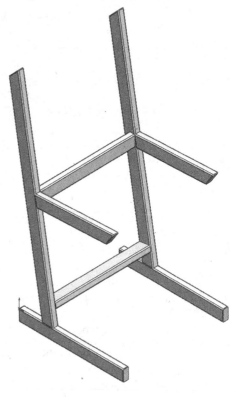

4. The designing of the profile ends
5. The modelling of a welded assembly
6. The cut list of profiles
7. The final model

10.3 Modelling in SolidWorks

SolidWorks has a special module for modelling constructions that are welded from various profiles. The welded construction is presented as a part that consists of more bodies (i.e., a multi-body part). The module has tools for the production of constructions from standard or alternative profiles that are joined on the basis of a sketch that presents the course of the axis of profiles in space. When connecting individual elements we can define the method of joining. We can also add various entities, like reinforcements, profile ends, welded assemblies, etc.

10.3.1 Creation of the System Structure: Skeleton

We define the courses of the profile axes with a sketch. Regarding the type of model, the sketch can be 2D or 3D. In the case of the chair construction (Fig. 10.12) we produce the sketch of the profile axis using the spatial sketch (*3D Sketch*) (Fig. 10.13). The welding skeleton or we can also say the construction system structure allows us to set cross-sections of the individual profiles in space so that the geometric centre of the profile is in the local coordinate system. In this case the local coordinate system runs on the structure line from the beginning to the end.

10.3.2 Profile Formating Through Skeleton

SolidWorks has a pre-set library of standard profiles. It is located in a default folder of profiles (***install_dir\lang\lang\weldment profiles***). We can also form an additional profile shape and save it in the appropriate location.

The software requires a specific shape and a structure of directories and files. The original folder contains one or more folders that represent the standard upon which the profiles are modelled. In this folder (***standard***) other subfolders are incorporated, which are named by the type of profile. In these subfolders the files of shapes of cross-sections for individual sizes are saved (Fig. 10.14).

In the case we do not find the desired profile in the pre-set library we can produce it with an additional process:

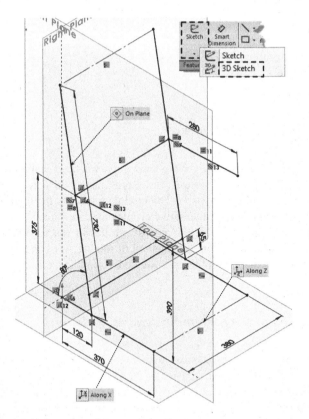

Fig. 10.13 3D sketch of creation of the system structure

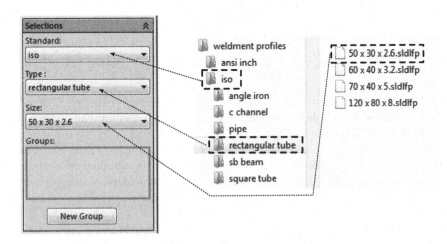

Fig. 10.14 Structure of files for defining the profile

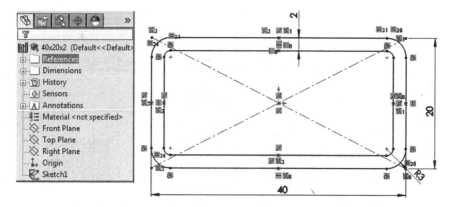

Fig. 10.15 Sketch of the profile that we wish to add to the base

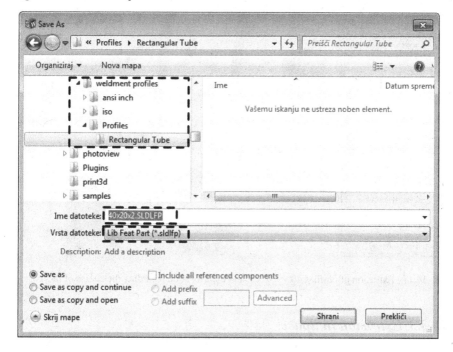

Fig. 10.16 Saving a new shape of the profile in the base

1. We create a new part.
2. We sketch the shape of the profile (we provide input nodes—the coordinate origin is default) (Fig. 10.15).
3. We first close the sketch and then we select it in the tree structure and save it in the appropriate folder (*Save As > Lib Feat Part*) (Fig. 10.16).

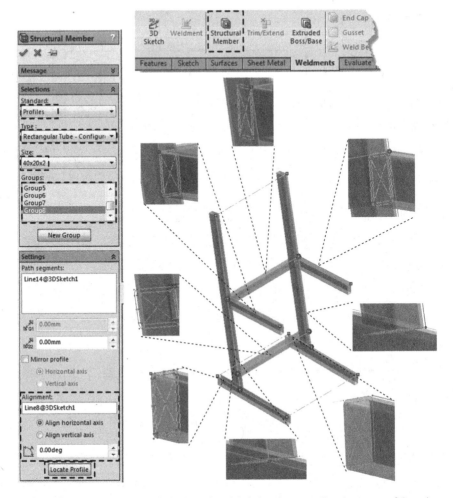

Fig. 10.17 Extrusion of profiles 40 × 20 × 2 and their location regarding the course of the axis

10.3.3 Extrusion in Space

To extrude profiles along the sketch (***Structural Member***) we first set the standard that determines the selected profile and later we define the type and the size of the profile (Fig. 10.17). With the formation of individual groups we set various positionings of the cross-section regarding individual elements of the sketch. Next, we select those segments of the basic sketch on which we wish to produce the extrusion. Later, we define the rotation and the location of the profile (***Locate Profile***).

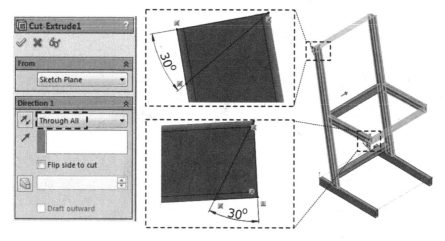

Fig. 10.18 Cut-off of the end of the profile at a specific angle

10.3.4 End-Forming Profiles

We transform the individual ends of the profiles into the desired end-forming profiles. For this reason we first cut (***Extruded Cut***) the profiles at a specific angle (Fig. 10.18). Especially with pipe constructions, it is important to close the opened parts of the profiles (***End Cap***). We can use common plastic caps or we can produce pads that we weld onto the profile (Fig. 10.19).

10.3.5 Welds Modelling

We usually do not model welded assemblies in the construction as a shape of **solid bodies**; instead, we present them as a mark in the model. The only exception is the corner weld (***Fillet Bead***) of a larger dimension, mostly for the purposes of presenting this joint in detail. In this case we also set the shape of the edges, which is important for the preparation of this welded joint.

SolidWorks uses a familiar presentation of welds modelling (***Weld Beads***), this enables the presentation of the shape of the welded joint, it calculates its length, etc (Fig. 10.20).

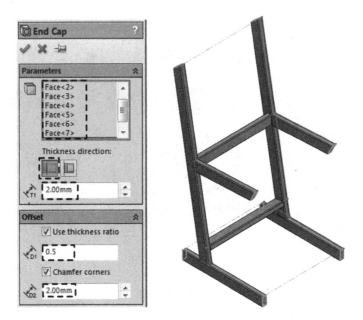

Fig. 10.19 Finishing profiles

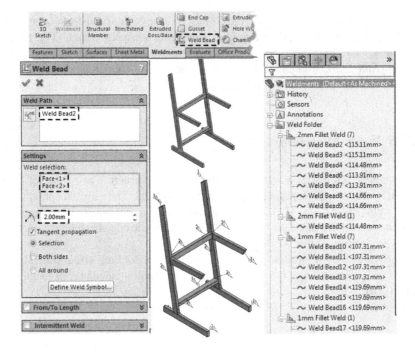

Fig. 10.20 Marking welded joints, their view and defining their lengths

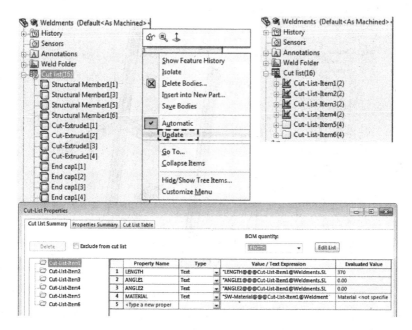

Fig. 10.21 Preparation of a cut list and the characteristics of the individual groups

10.3.6 Cut List of Individual Parts of a Welded Construction

The main element of the technical documentation of the welded construction is its cut list of individual parts (**Weldment Cut List**), where the shapes and quantities of the component parts of the welded assembly are defined. First, we **Update** the cut list in the feature manager. Next, we have groups of individual equal elements of the construction, to which we give names. Later, we can define the characteristics of each element (Fig. 10.21).

10.3.7 Final Model of the Welded Construction of a Chair

Finally, we determine the material (**Edit Material**) and read the mass of the model of the welded construction of the chair (Fig. 10.22).

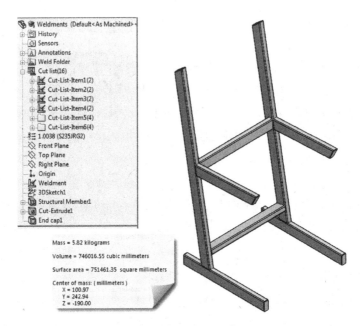

Fig. 10.22 Final model of the frame of the chair with its structure of features and mass characteristics

10.4 Modelling in NX

10.4.1 Creation of the System Structure: Skeleton

We begin the modelling by choosing the XC–ZC plane of the main coordinate system on which we sketch the main structure lines of the construction. To sketch we use the command *Sketch* and we produce the side profile of the chair (Fig. 10.23).

To define the spatial shape of the steel construction of the chair we use the command for forming a new plane that is parallel. The plane is called an auxiliary plane. This is performed using the command *Feature > Datum Plane* that sets the auxiliary plane at a distance of 380 mm from the basic plane (Fig. 10.24).

We activate the new sketch in the auxiliary plane using the command *Sketch/ Project Curve* and project the lines from the first sketch. We then finish the sketch on the auxiliary plane (*Finish Sketch*). In the next step we produce the connective structural lines between the two sketches (*Curve > Line*) and connect the junctures between the sketches in both planes (Fig. 10.25). The connective structural lines are geometric entities that are not part of the sketches, but are independently defined in the data structure, similar to the individual features.

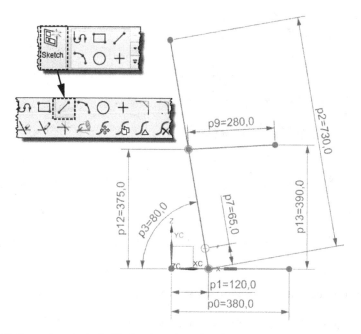

Fig. 10.23 Formation of a basic 2D sketch on the main XC–ZC plane

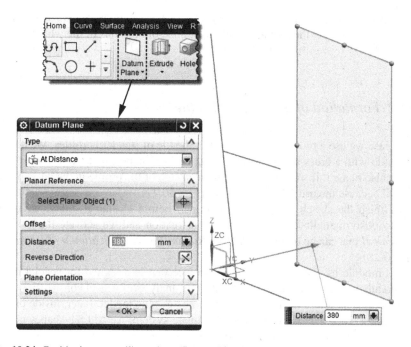

Fig. 10.24 Positioning an auxiliary plane (*Datum Plane*) that is parallel to the X–Y plane and at a distance of 380 mm

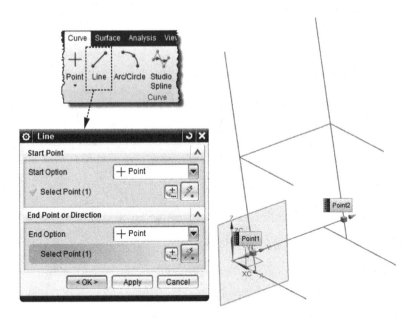

Fig. 10.25 Spatial connective structural lines (*Curve/Line*)

The positions of the profiles and the points of the junctions that we will define hereinafter must meet the topology and the defined dimensions of the whole model (Fig. 10.26).

10.4.2 Formation of a Standard Profile

In our case we use a profile that is not yet in the file of standard profiles. We define the profile with a cross-section sketch using the command *File > New > Part* and we can also name it if we wish. We form the sketch of the profile in the basic plane XC–YC because the origin is in that plane (Fig. 10.27).

We finish the sketch and save the new part before closing it. Later, we open the previously modelled skeleton of the construction and we activate the module *Mechanical* that can be found in the *Main Ribbon Bar > Application > Mechanical* (Fig. 10.28).

This module has an integrated user interface that is designed for special uses. The module uses the known concept of a static analysis of a steel construction. We first prepare a structure of the construction, i.e., a static model. Later, inside of the static model, with regard to the local coordinate system, we add the shapes of the cross-sections of the profiles that we form as connective profiles from one to another junction (Fig. 10.29).

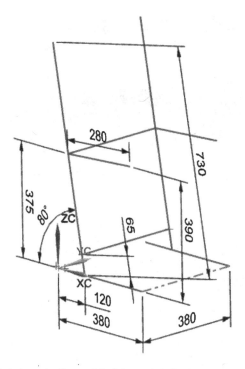

Fig. 10.26 Produced skeleton for the model of the steel chair (the static model of the chair)

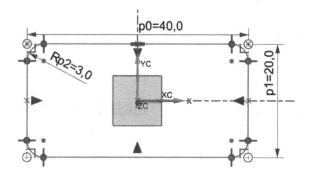

Fig. 10.27 Sketch of the profile in the basic plane XC–YC

To define the structure of the static model in this new environment we first need to define the previously produced profile. To accomplish this we need the information below:

- The junctions—*Anchors*
- The detailed shape of the profile (*Simple, Detail*)

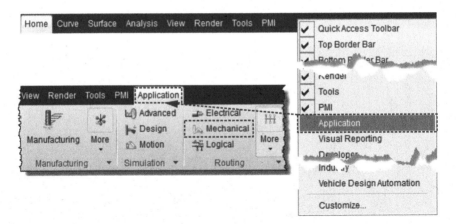

Fig. 10.28   Activation of a special module (*Routing Mechanical*)

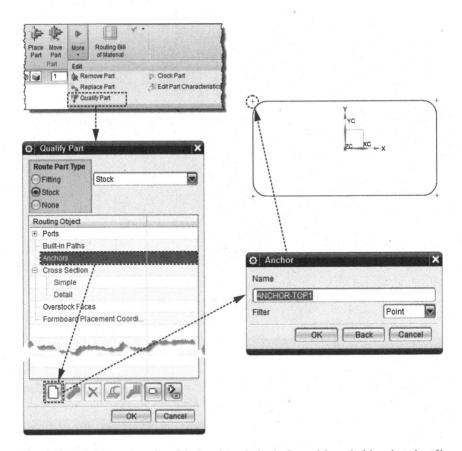

Fig. 10.29   Definition and naming of the junctions, the beginning and the end of the selected profile

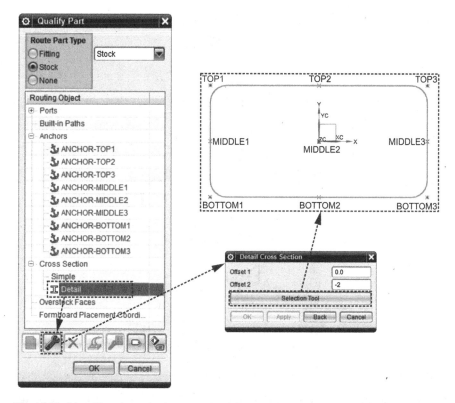

Fig. 10.30 List of junctions (*Anchors*) and the definition of the detailed shape of the profile

We select the tab *Stock* using the command *Qualify Part* where we can define a collection sheet and we continue on the tree structure (Fig. 10.30). Later, we define the junctions or anchors as the start and end points are named in the software. The points of the junctions are the points of the beginning and the end of the structure lines of the static model. We define nine points and we name them (Fig. 10.30). Now the static model is added in the database.

The settings in the menu *Qualify Part/Cross Section* provide the display mode for the profiles in space. The specification of the profiles is important for complex constructions or if the mechanical equipment is limited. In our case we can show a simplified shape of the profile (*Simple*), which is less demanding for the topologic display than the actual model defined in detail (*Detail*) (Fig. 10.31).

10.4.3 Extrusion of Profiles in a Skeleton Structure

Later, we try to present the entire steel construction in the most realistic way. Employing the extrusion we use certain cross-sections of the profiles in junctions from the

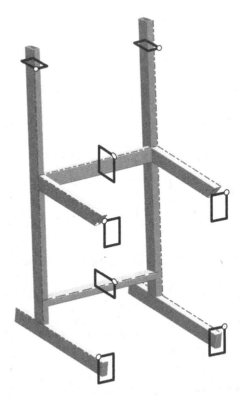

Fig. 10.31 Schematically presented positions of the cross-sections of the individual profile regarding the skeleton

beginning to the end (***Create Linear Path***). This feature enables the extrusion from the beginning to the end of the profile with previously defined junctions of the entire skeleton of the steel construction (Fig. 10.32).

We must emphasize that the orientation of the cross-sections in the coordinate system is also important in this case. We have to be careful that the start points are in the same start of the same quadrant. This is a typical error for the first use of the software or the first modelling experience. We start modelling using the command ***Stock/Orientation*** that defines the orientation of the cross-section. We first choose the appropriate file that contains the profile (***Specify Stock***) and later we determine the start point in the profile definition that is the ***Anchor*** (Fig. 10.33). To ensure the correct position of the cross lines we also use the function of a rotation angle to the line of the structure skeleton (***Rotation***), which precisely defines the position of the cross-section in the origin of the local quadrant (Fig. 10.34).

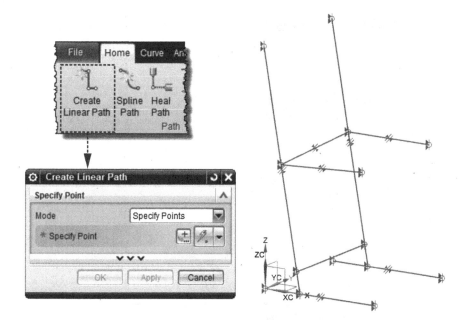

Fig. 10.32 Definition of the direction and the length of the extrusions (***Create Linear Path***)

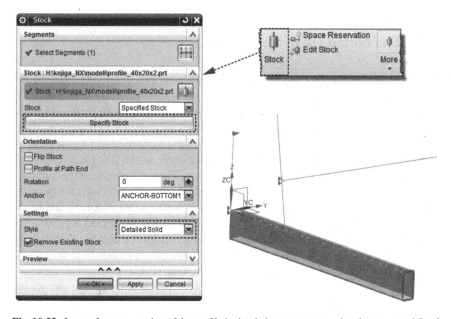

Fig. 10.33 Input of a cross-section of the profile in the skeleton structure using the command ***Stock***

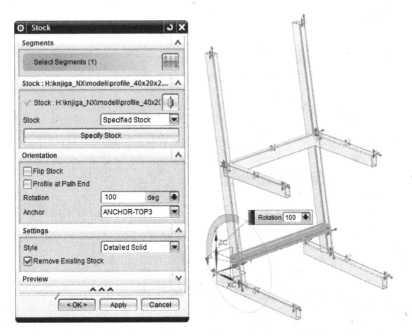

Fig. 10.34 Orientation of the cross-sections and the positions of the profiles (the beginning and the end) on the skeleton

10.4.4 End Formation of the Profiles

It is logical that the joints of the static model of the steel construction are the origins for a detailed formation of the construction of the nodes. To easily form nodes of the steel construction we need to adjust the individual lengths of the profiles in parts of the starting and the final lengths of the profiles (Fig. 10.36). We perform that with the feature **Replace Face**, which is located in the basic module (**Application > Modeling > Synchronous Modeling > Replace Face**). To use this feature we first need to select the surface that we wish to substitute. Then, we choose a new surface. The program executes this operation so that it substitutes the selected surface with the new surface and automatically adjusts the other surfaces. Now we have an object that has an entirely closed volume and the steel construction is defined (Fig. 10.35).

We produce covers for the legs of the chair using extrusion. Later, we will present a few examples of the modelling of the end-formation profiles into a final shape and the correlation with the technological production (Fig. 10.37).

In some cases the modelling concept that uses special features like the set of commands for the rapid modelling of welded assemblies is not consistent with the final shape of the model. In the case shown in Fig. 10.38 we set a sheet metal on the perpendicular on the opening of the beam so that it is in the vertical direction on both sides, 1 mm above the size of the opening. In the manufacturing process we weld it

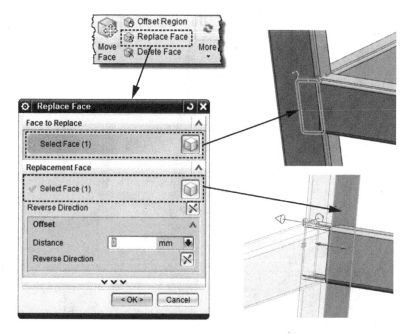

Fig. 10.35 Adjusting the lengths of the profiles in the junctions of the structure (***Replace Face***)

on to the beam to prevent it from dropping into the opening because it is larger than 1 mm.

First, we compare the various uses of the feature in the module *Weld Assistant* with the produced geometry of the welding (procedurally used, already-know features) (Fig. 10.39).

In the module *Weld Assistant* there is an option for modelling and symbolic marking in the model in 3D space.

We need to remodel the original shape of the sheet metal for the cover by adding the material (***Extrude***) (Fig. 10.40). We begin with the fact that we produce the cover that is in outer dimensions identical to the extended plane of the inner cross-section of the steel profile. If we were to manufacture this cover, it would fall into the opening, and we cannot lay it on the welding place. We need to support the cover on the edges of the opening of the steel profile. So, we extend the cover on two sides by 1 mm (2 mm is the maximum) regarding the size of the cover. With the extended sides we can lay the cover on the opening and we can weld it to the profile of the steel construction with two fixing welded assemblies. Now it is possible to weld (Fig. 10.41). The detail we have presented shows the importance of accurate modelling for the definition of all the dimensions of the elements of the steel construction (Fig. 10.42).

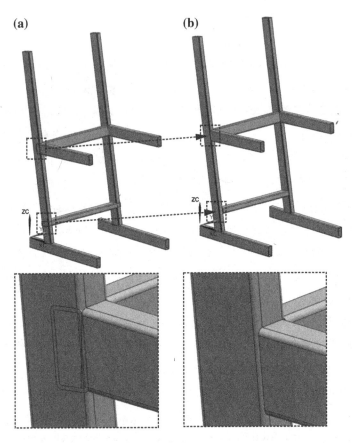

Fig. 10.36 Additional view of adjusting the lengths of the profiles in the junctions of the skeleton: before adjusting (**a**) and after adjusting (**b**)

Finally, we model welded assemblies on other places of the joint (Fig. 10.43). We model the welds in detail when they are weighed down, because in these cases they need to be accurately defined and in accordance with their function. In other cases we can use the module *Weld Assistant*, but the constructor always has to have in mind that his or her job is to accurately model the manufacturing process of the product (Fig. 10.44). Accurate modelling is also needed for a calculation of the mass of the product and indirectly for the masses of the welded assemblies. The precise geometry is necessary for potential numerical simulations of the stress-deformation analyses. If these details are not needed, it is enough to just mark the welded assemblies with symbols (Fig. 10.45).

Product Manufacturing Information (PMI) is a tool for marking and dimensioning the parameters and dimensions in 3D space (analogous to the dimensioning in 2D sketches) (Fig. 10.46).

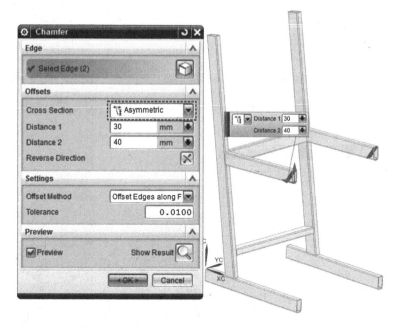

Fig. 10.37 Forming end-formation profiles in all the junctions of the steel construction with the feature *Chamfer*

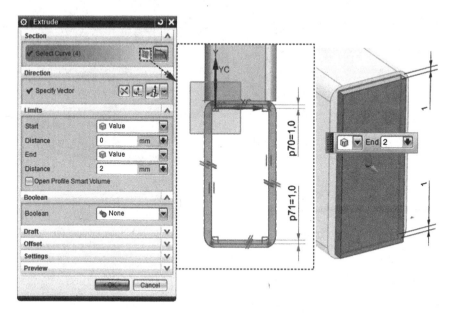

Fig. 10.38 The basic principle of the modelling that is equal to the manufacturing technology

Fig. 10.39 Interface module for the modelling of welded assemblies

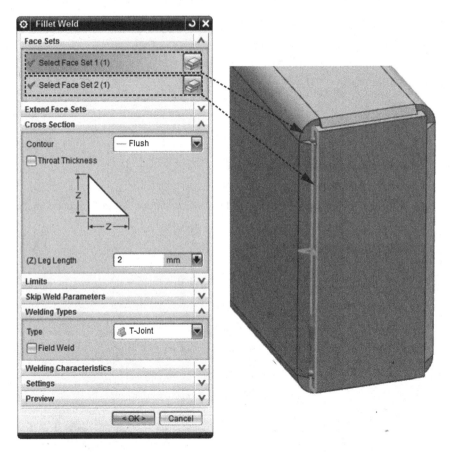

Fig. 10.40 Highlighted sharp edges of a sheet metal that are modelled in the module for welded assemblies do not form the correct shape

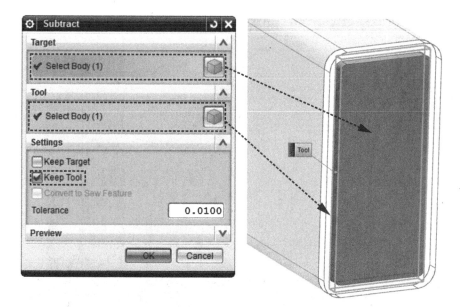

Fig. 10.41 Additional extrusion of the model of the cover that is produced to adjust the geometry of the elements of the steel construction to the actual details in the welding process

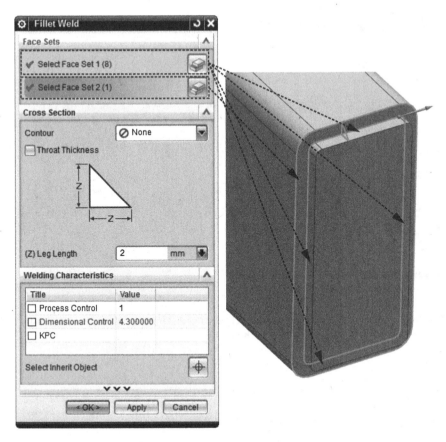

Fig. 10.42   Modelling a welded assembly between the cover and the end of the profile of the steel construction using the command *Fillet Weld*

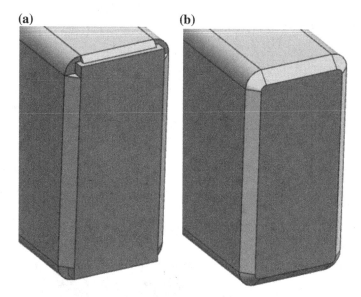

Fig. 10.43 Comparison of both processes and the resulting shape of the final welded assemblies. The left model (**a**) is an example of the use of the module *Weld Assistant*; the right model (**b**) is the modelled welded assembly in the actual shape

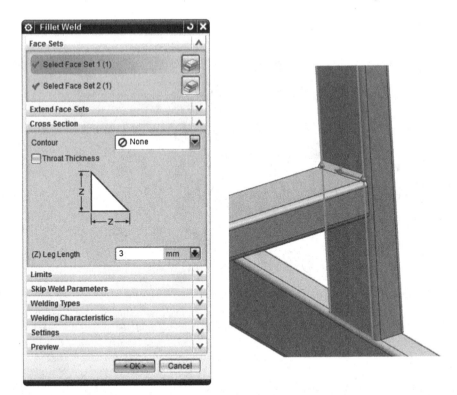

Fig. 10.44 Example of modelling a triangle welded assembly (*Fillet Weld*) on the rest of the parts of the welded construction

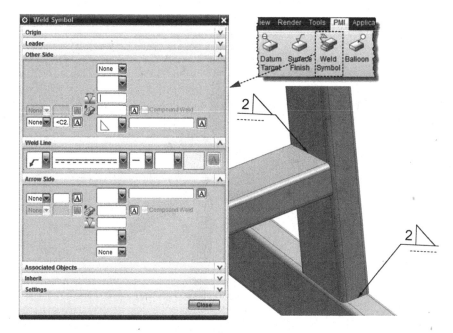

Fig. 10.45 Marking welded assemblies (*PMI > Weld symbol*)

Fig. 10.46 Final model of the welded steel chair

10.5 Examples

Figures 10.47, 10.48, 10.49 and 10.50 present few examples of welds.

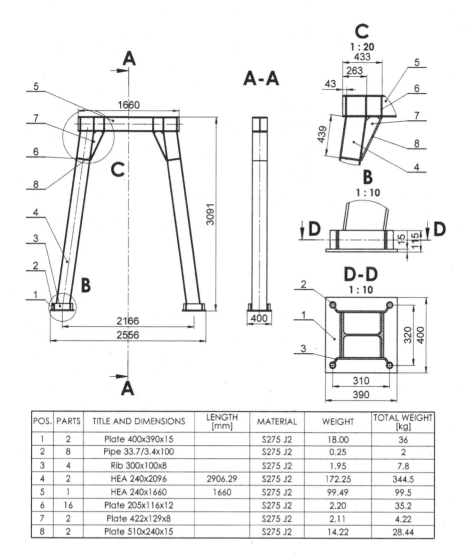

POS.	PARTS	TITLE AND DIMENSIONS	LENGTH [mm]	MATERIAL	WEIGHT	TOTAL WEIGHT [kg]
1	2	Plate 400x390x15		S275 J2	18.00	36
2	8	Pipe 33.7/3.4x100		S275 J2	0.25	2
3	4	Rib 300x100x8		S275 J2	1.95	7.8
4	2	HEA 240x2096	2906.29	S275 J2	172.25	344.5
5	1	HEA 240x1660	1660	S275 J2	99.49	99.5
6	16	Plate 205x116x12		S275 J2	2.20	35.2
7	2	Plate 422x129x8		S275 J2	2.11	4.22
8	2	Plate 510x240x15		S275 J2	14.22	28.44

Fig. 10.47 Slope lift pilar

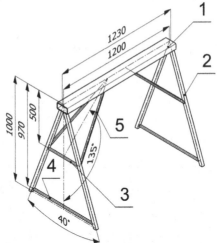

Pos.	Qty.	Title and dimensions	Length [mm]
1	1	Furniture tube ⌷ 120 x 60 x 5	1230
2	4	Furniture tube ⌷ 30 x 30 x 3	1012
3	2	Furniture tube ⌷ 15 x 15 x 1,5	338
4	2	Furniture tube ⌷ 15 x 15 x 1,5	680
5	4	Furniture tube ⌷ 15 x 15 x 1,5	638

Fig. 10.48 Support for table H = 1,000 mm

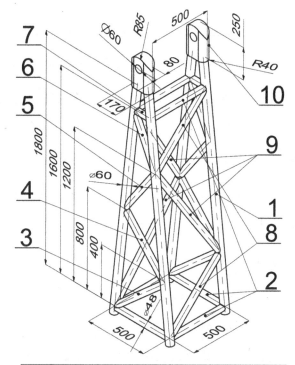

Pos.	Qty.	Title and dimensions	Length [mm]
1	4	tube Ø60x3	1865
2	6	tube Ø48x2,5	500
3	2	tube Ø48x2,5	606
4	2	tube Ø48x2,5	543
5	2	tube Ø48x2,5	487
6	2	tube Ø48x2,5	442
7	2	tube Ø48x2,5	145
8	2	tube Ø48x2,5	642
9	3	tube Ø48x2,5	948
10	2	top part 250x170x80	

Fig. 10.49 Support H = 1,800 mm

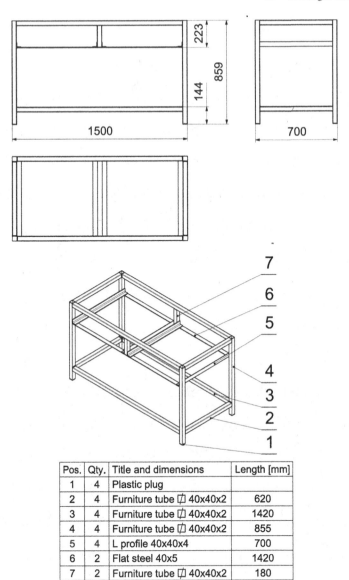

Pos.	Qty.	Title and dimensions	Length [mm]
1	4	Plastic plug	
2	4	Furniture tube ⌷ 40x40x2	620
3	4	Furniture tube ⌷ 40x40x2	1420
4	4	Furniture tube ⌷ 40x40x2	855
5	4	L profile 40x40x4	700
6	2	Flat steel 40x5	1420
7	2	Furniture tube ⌷ 40x40x2	180

Fig. 10.50 Table 859 × 700 … 1.500 mm

Chapter 11
Sheet-Metal Bending

Abstract Allowing a high degree of repeatability and being relatively cheap in terms of their function, sheet-metal products nowadays enjoy a significantly advantageous position in terms of their use. Besides edge and fold details, sheet-metal products require some important specifics for defining the unfolded sheet metal. In this chapter attention is drawn to the details that the user should be careful about when using automated default settings, both for bending radii and for different materials (elastic module E). Two methods for defining the sheet metal part: (1) from a cover or (2) from a solid material are pre-set.

A special technology for the manufacturing of sheet-metal products requires a particular method for the modelling and setting the details. Different features are used for both the manufacturing technology and the modelling. There are different technologies available for changing the shape of the sheet metal: cutting, chopping, punching, arching, bending, pulling, etc. Each of them requires a specific manufacturing process and model preparation.

There are two basic techniques for the modelling of sheet-metal products. (1) Creating a model from flat sheet metal, leading to the final shape through various bending, punching and reshaping procedures. (2) Forming a product as a solid full-shape model and then transforming it into a sheet-metal product (Fig. 11.1). In the latter case, the product's outside shapes are used in such a way that it has no closed shapes on the inside. The open outside shapes allow shearing the sheet metal with a cutter. Cutting the inside closed shapes usually requires laser, plasma or water-jet cutting.

This exercise creates familiarity with making sheet-metal products by changing the shape of the material, mostly by folding flat sheet metal into the end product. For a better understanding of the procedures, we will first define a starting global shape that will then have various reinforcements and panels added on the way to a complete model with a final shape.

The modelling of sheet-metal products includes two steps: (1) creating a cover for the unfolded sheet metal and (2) the cutting plan. Modern technology normally

(a)   **(b)**

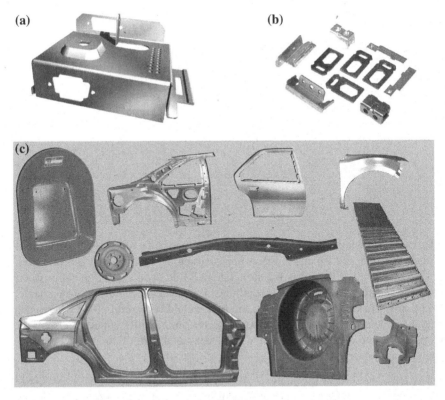

Fig. 11.1 Sheet-metal products (*source* http://www.indiamart.com/, http://www.metexauto.comm, http://www.atlastool.com): housing part of control unit (**a**), elements for doors and windows (**b**) and automotive sheet metals parts (**c**)

employees laser or plasma cutting. When defining the final shape, consider the cutting width (e.g., between 0.12 and 0.24 mm for laser cutting), which is the basis for creating a DXF file for a direct data transfer to the cutting-machine controller. Having defined all the above-mentioned steps and technological conditions, proceed to creating the corresponding bending documentation that generally includes the unfolded surface and the final bent shape, representing a product or an intermediate product to be welded or used directly. The end product is normally presented as a 3D model on the same plan, together with the control dimensions.

11.1 Manufacturing Technology and Use

Special presses are the most frequently used machines to bend and cut sheet metal (Fig. 11.2). More complex shapes require special, multipart tools. On the bending line there is a cutting mechanism or a profile tool with grooves for rectangular or

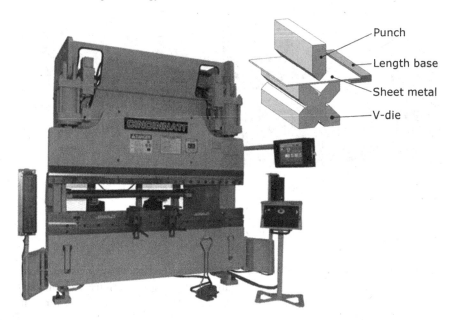

Punch
Length base
Sheet metal
V-die

Fig. 11.2 Computer-controlled sheet-metal bending press and a bending tool

other types of reshaping. This reshaping is linear up to the maximum tool length. Note that there are three classes of bending technologies, depending on the thickness of the sheet metal: (1) thin sheet metal up to 1.5 mm (2 mm), (2) up to 4 mm (6 mm) and (3) up to 15 mm (20 mm). In comparison to other methods, like cutting, bending remains the cheapest method for reshaping.

11.2 Modelling Sheet-Metal Products

Sheet-metal products are modelled with a specialized software interface. In principle, it is a derivative of a number of basic features, such as extrusion or revolving. It is adapted for use and allows the smooth modelling of sheet-metal products. The key parameter is the sheet-metal thickness. It is a constant and represents the starting point for all the other values that are important for defining a detailed shape. Because specifying detailed shapes depends on a material's properties it is always necessary to specify the material before specifying the flattened sheet metal and other details (Fig. 11.3).

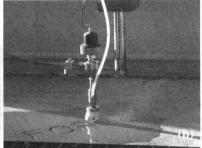

Fig. 11.3 Sheet-metal cutting: laser (**a**) and water jet (**b**)

11.2.1 Definition of the Material's Parameters

When specifying the sheet-metal shape, it is necessary to first define the sheet-metal thickness and then the material. The details of the shape's various technological patterns should be specified in a special menu. Modelers usually have settings in their main menus.

To make the use of the input conditions clearer, some typical examples of the key parameters' entry are presented below.

- **Material thickness**, i.e., the sheet-metal thickness should be set before the modelling begins and should remain the same throughout the modelling. This can change only if you return to the beginning.
- The **bend radius** depends on the sheet-metal thickness in the sense that it takes account of a certain factor (e.g., 0.8–2.0 × d). The most frequent ratio is 1.0, i.e., the bend radius equals the sheet-metal thickness radius. A smaller radius results in a rapid increase of the bending forces and particularly the residual stress in a material due to increased plastic deformations. It causes increased return deformations after bending—the so-called "spring-back" effect. The setting can be individually reset for certain features.
- **Bend relief depth/width** are two design parameters that define the detailed shape of the bend relief on sheet metal (Fig. 11.4).
- **Bend Allowance Formula**—an example of a default setting: (Radius+(Thickness × 0.44)) × rad(Angle)

11.2.2 Methods of Modelling Sheet-Metal Products

When modelling a sheet-metal product, there are generally two typical procedures:

1. Creating products' covers. It is essential to start with the function of a cover as a solid body (Sheet metal from a solid). The volume is then defined as the body's

Fig. 11.4 The bend relief
depth/width can be pre-set

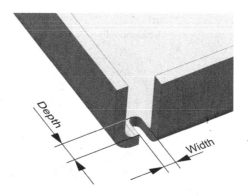

lateral surface, which is then developed into a flat lateral surface. To do this, there are usually special features available for mapping the geometry into a suitable shape of a sheet-metal product. The software takes account of the shape of the bent edges, deformations, bend reliefs, etc (Fig. 11.5). The model in its final shape is so adapted that the sheet metal can be flattened on a plane.
2. This procedure starts from a base sketch, and the final shape is modelled step-by-step. Having finished the modelling, you should also create the flattened sheet metal, which is always the final and—most importantly—the key information to prepare sheet-metal cutting, regardless of the cutting technology to be used. Specific cutting technologies always require additional work on lengths and detailed shapes in order to be able to execute accurate dimensions during the cutting itself.

Figure 11.6 shows the individual steps necessary to create the support (Fig. 11.6c), reshaped with the (*Flange*) command from the plate Fig. 11.6a into b. Cutting out a rectangular shape and simultaneously reshaping the basic shape is executed with a special (*Cut out*) command, shown in Fig. 11.6c. The procedure ends by unfolding the sheet metal (Fig. 11.6d). In this case, we would like to present cut out modelling, which is different from manufacturing practice. In manufacturing, the sheet metal is first cut out after unfolding the shape (Fig. 11.6d), followed by bending three times (Fig. 11.6b) and the final shape is the same (Fig. 11.6c).

Modelling should therefore be applied step-by-step, in a logical and careful way.

11.2.3 Features and Settings

- **Creating Base Tab**
  Using this feature and extrusion into space, which is hidden, you can create a solid body from a 2D sketch. The body of a model allows defining the lateral surface of the sheet metal (Fig. 11.6a).

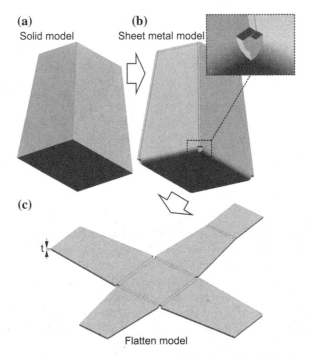

(a)
Solid model

(b)
Sheet metal model

(c)

t

Flatten model

Fig. 11.5 Modelling a sheet-metal product from a solid. Solid model (**a**). Sheet metal model (**b**)

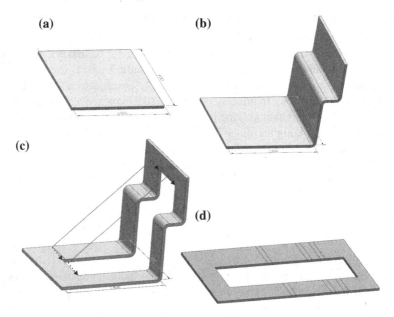

(a)

(b)

(c)

(d)

Fig. 11.6 Step-by-step modelling of a product from a base sketch, following steps (**a**)–(**d**) to the final shape

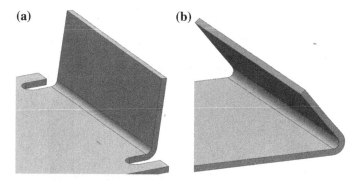

Fig. 11.7 Creating flanges and a flap with the (*Flange*) command. Partly bending (**a**). Total line bending (**b**)

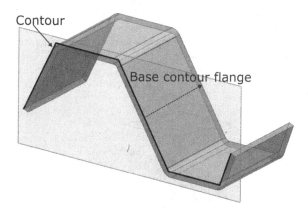

Fig. 11.8 Modelling bent sheet metal from a base sketch (*Base Flange*)

- **Hem/Flange/Base Flange**
  This feature is useful for reinforcing the edges and is typical for designing sheet-metal products. Each edge, whose length is more than 10 times the thickness of the sheet metal, should be reinforced with a flap. It usually considers the technological laws of manufacturing (reliefs, radii, etc.) of a 3D model (Fig. 11.8).
- **Creating transitions** in sheet-metal products (*Lofted Flange*)
  Modern modelers have active features to automatically determine the transitions between two different cross-sections. The software adjusts the area of the lines of the bending and simultaneously considers the bending deformation (Fig. 11.9).
- **Cut-out**
  Modelling cut outs of different shapes in sheet metal (inside corners, etc.) is similar to removing material in extrusion (Fig. 11.10). In this case, the sheet-metal bending characteristics and a direct projection on the surface of the model are considered.

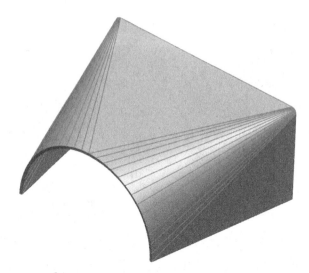

Fig. 11.9 Creating transitions between different cross-sections, while taking account of the technological characteristics of the manufacturing

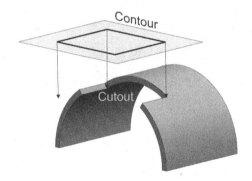

Fig. 11.10 Modelling a cut out, using a sketch projection on the surface where the cut-out feature is used

11.2.4 Setting Bent Edges

- **Reference length** (Fig. 11.11)
- **Setting the bend radius on the edges**

Modelling flaps (*Flange*) is critical for sheet-metal products. When choosing the procedure, you should select the dimensions that can be later used for the execution of the final model or flattened sheet metal. In view of the importance of a particular shape, you should decide whether to use the outside or the inside length when determining the inside or—exceptionally—the outside edge (Fig. 11.12).

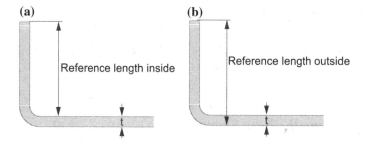

Fig. 11.11 Reference length and its consideration for specifying the flap length, depending on the sheet-metal thickness: reference length inside (**a**) and reference length outside (**b**)

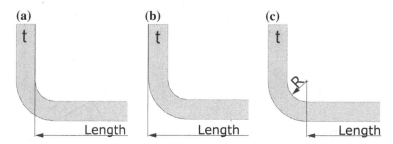

Fig. 11.12 Reference length and its consideration for specifying the flap length, depending on the sheet-metal thickness: inside length (**a**), outside length (**b**) and flat length (**c**)

11.2.5 Modelling Sheet-Metal Products

Modellers differ from one another in terms of modelling sheet-metal products. They all include the criteria of plastic deformation, vital for determining flat surfaces. Specific details also depend on the design of the tools that are used for creating the corners. To define the cut-outs or to add special shapes, apply the familiar procedures from previous chapters. Because these procedures vary from one modeller to another, it is good to check both the SW and NX presentations and then make a judgement about the differences that make using one or other software easier.

An important point is about using a particular technology to create corners and reinforcements on the edges.

An example of a casing model is presented in Fig. 11.13.

The modelling procedure can be broken down into the following steps:

1. Creating the base shape
2. Adding the boundary faces
3. Completing the shape
4. Unfolding the sheet metal and cutting
5. Finishing the model.

11.3 Modelling in SolidWorks

To model a product in SolidWorks, shown in Fig. 11.13, activate the set of commands for modelling sheet-metal parts (***Sheet Metal Parts***) (Fig. 11.14).

11.3.1 Creating the Base Shape

The base shape is usually drawn on a base plane. In our case we use a side drawing plane, which is at the same time also a symmetrical plane. The base shape can be submitted in the form of an open contour, representing the outside surface of bent sheet metal. Next to the open contour, automatically draw the inside contour, which has the thickness of the sheet metal, and then mark this surface as the basic one.

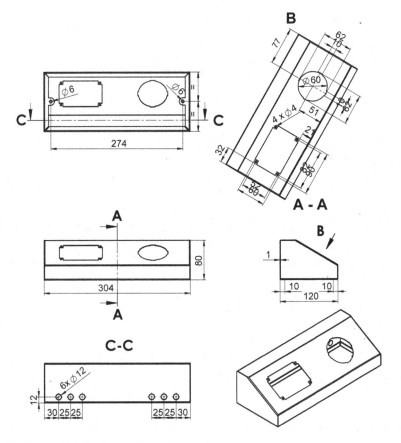

Fig. 11.13 A model of a sheet-metal casing

Fig. 11.14 The set of commands for modelling sheet-metal parts

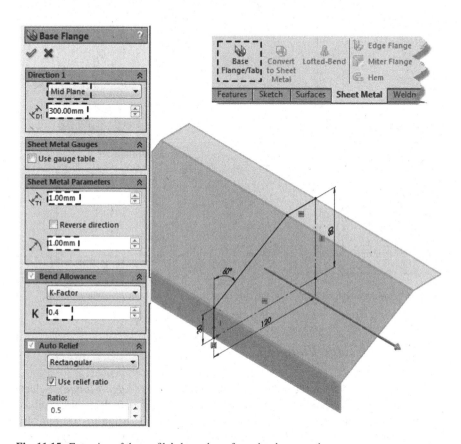

Fig. 11.15 Extrusion of the profile's base shape from the sheet metal

Once the basic surface has been set, the extrusion can be executed (***Base Flange***) (Fig. 11.15), specifying the extrusion parameters, the sheet metal thickness, the inside bending radius and the K-factor, describing the behaviour of the material during the bending. The K-factor represents the deformability of the material in a plastic area, i.e., its plastic elongation in the bending area, which extends the length and indirectly marks the thinning in the bending area.

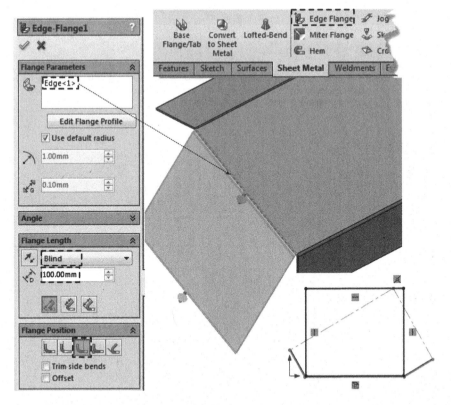

Fig. 11.16 Adding an edge flap to close the side face

11.3.2 Adding Edge Flanges

The casing is closed from the side by adding edge flanges (*Edge Flange*). Select an appropriate (the longest) edge and specify the tab parameters. It is important to select the proper tab position. Set the tab length for the extrusion to 100 mm. In the sketch (*Edit Sketch*), copy it onto the base sketch with the (*Coincident*) relation (Fig. 11.16).

The edges of the base profile are usually different from tab edges, resulting in empty spaces between the tab and the other profile edges, which need to be filled. This operation is executed with the (*Closed Corner*) command, yielding a contact between the individual faces of the cover. Namely, the air vents or the overlapping should be reduced as clearly as possible, using simple shapes. The corner-closing operation is shown in Fig. 11.17.

When closing the corners, excessive parts of the faces can remain and need to be cut off using the (*Cut-Extrude*) command. Cutting is usually focused on the size of the cutting and the material thickness. The (*Cut-Extrude*) function is therefore used

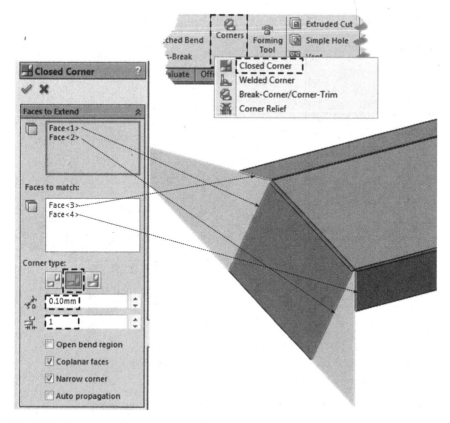

Fig. 11.17 Closing the corners of the side face

to correct detailed shapes in order to be able to form corners in the lines and not to prevent flattening or welded joints. Use the cutting shape and the size from the existing geometry with the *Convert Entities* command. Next, link the cutting depths to the material thickness (Fig. 11.18).

Create the panel on the other side in the same manner. Because the product is symmetrical across the side plane, mirroring (*Mirror*) can also be used. Pay attention to the selection of the mirroring features (Fig. 11.19).

Create the back face in a similar way to the side panel by adding the edge flaps (Fig. 11.20).

Add a reinforcement on the bottom part, which significantly strengthens the entire cover. This is executed by adding edge flaps (*Edge Flange*). For technological reasons, the edge flap in the free part should not be less than 8–10 times the sheet-metal thickness, or it can be limited by the tool length. In our case, we set the value at 10 mm (Fig. 11.21).

In the middle of the left or right panel, add two tabs on the lower reinforcement, which is intended for the positioning of the piece. In doing so, make use of the

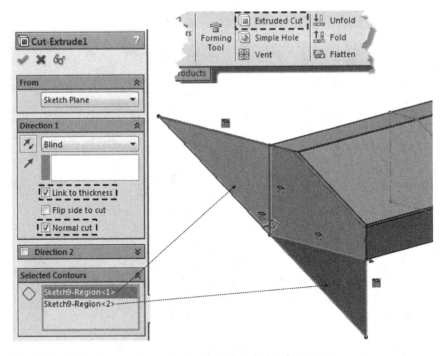

Fig. 11.18 Cutting the excessive part of the face after closing the corners

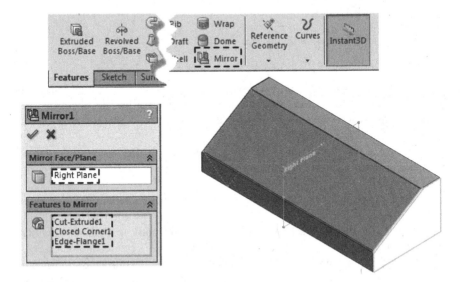

Fig. 11.19 Mirroring the side panel across the right (*side*) plane

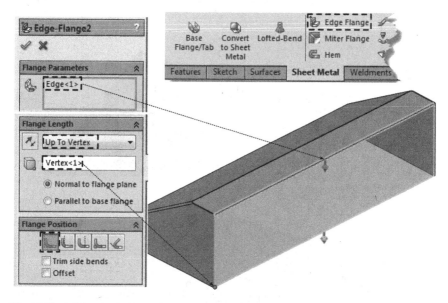

Fig. 11.20 Adding the back panel

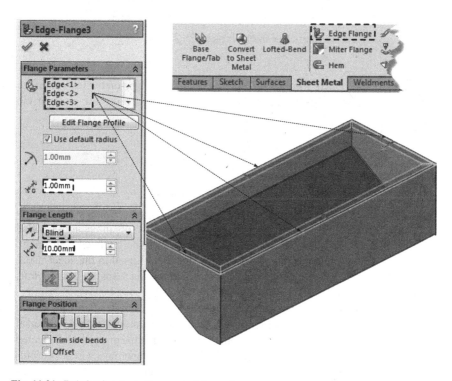

Fig. 11.21 Reinforcing the bottom part of the casing

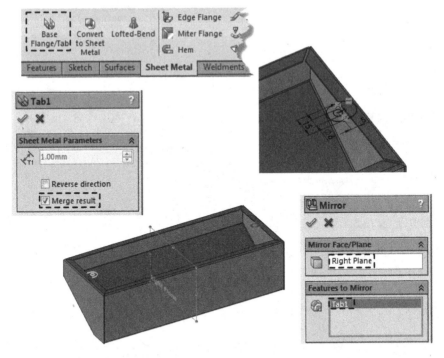

Fig. 11.22   Adding a positioning tab on the lower face

material adding the (**Base Flange/Tab**) function. Draw a shape on the lower surface and extrude it. Create the tab on the opposite side by mirroring the existing tab (Fig. 11.22).

11.3.3 Completing the Shape

In order to complete the shape, create the remaining elements that are required to fulfil the function and the shape. To install instruments, two openings are provided on the front face. The shape of the openings depends on the intended instrument. Create the opening by means of extrusion, with the depth linked to the sheet-metal thickness. Due to the relatively complex sketch, it is important to mark all the regions that need to be cut out (Fig. 11.23).

On the back panel there are three holes with an appropriate diameter, which are intended for electrical feed-through connections. Because this is a universal regulation panel, allowing connections from the left or right, the openings are created as cut-outs, making it possible to remove only the required openings. Create the cut-out profile as a simple extrusion. By means of a linear pattern, create three openings across 25 mm, and transfer them by mirroring onto the other side (Fig. 11.24).

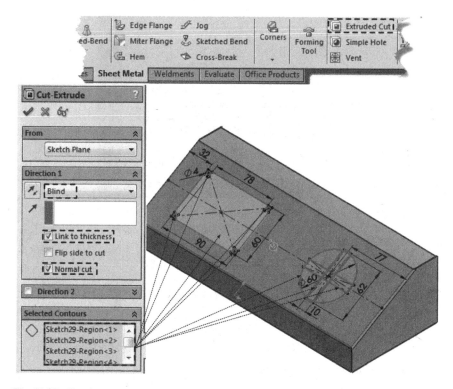

Fig. 11.23 Creating cut-outs on the front face by means of extrusion

The details of each product provide feasibility or some higher degree of techno-logicality. In our case, roundings on the left and right side panels need to be created (**Break-Corner**). For folding reasons, use the same radii as the rounding on the outside surface (Fig. 11.25).

11.3.4 Unfolding the Sheet Metal and Cutting

The unfolding of the sheet metal is an important step in the design of sheet-metal products. Individual edges/panels can be unfolded for creating or controlling special shapes. Unfolding the entire surface is mostly intended for producing the cutting documentation.

A single fold is executed by means of the unfolding function (**Unfold**). To do this, specify a fixed face and the edges to be flattened. Similar to unfolding, there is also a re-folding function (**Fold**). However, it only works when a shape has already been unfolded (Fig. 11.26).

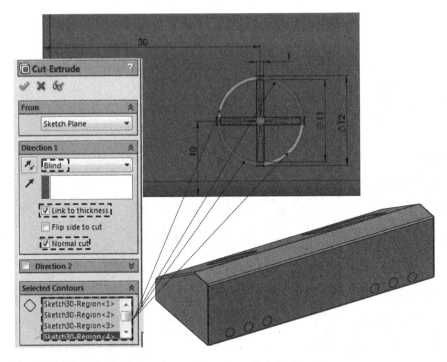

Fig. 11.24 Modelling feed-through cut-outs on the casing's back panel

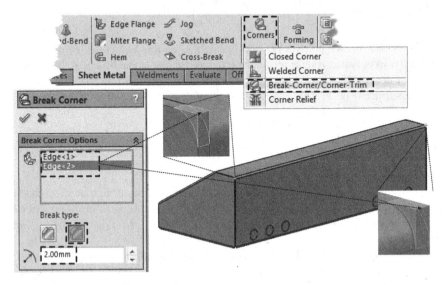

Fig. 11.25 Rounding the edge of the side panels

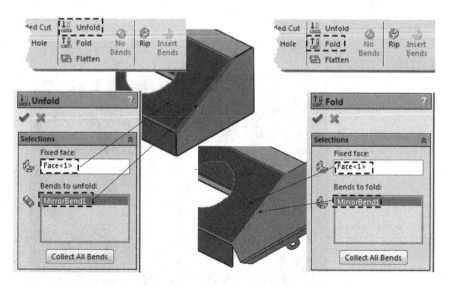

Fig. 11.26 Unfolding and re-folding a bend

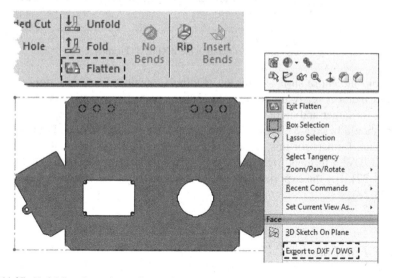

Fig. 11.27 Unfolding the entire surface and exporting to the DXF or DWG format

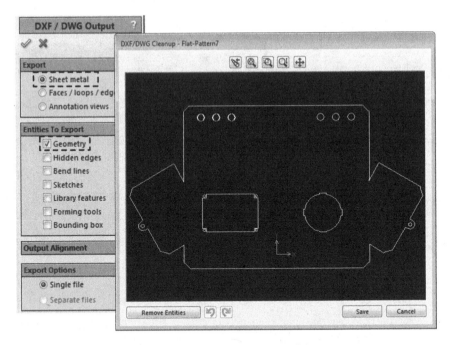

Fig. 11.28   Exporting the unfolded sheet metal in the DXF/DWG formats

Unfold the entire surface with the (*Flatten*) function. Such an unfolded piece of sheet metal is ready for forming the cutting plan. Laser cutting is particularly suitable for larger and more complex shapes. For this purpose, an appropriate document should be created, specifying the exact cutting path. DXF and DWG are two established formats for this purpose. To save an unfolded piece of sheet metal in the DXF/DWG formats, right click the surface and select **Export to DXF/DWG** (Fig. 11.27). For the export, select only the geometry of the unfolded sheet metal (Fig. 11.28).

11.3.5 Final model

Figure 11.29 shows a complete model of a sheet-metal housing with the corresponding features structure and its mass characteristics.

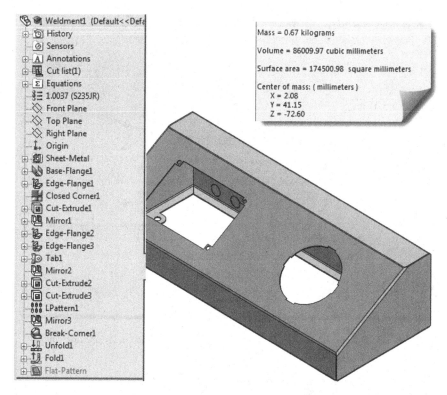

Fig. 11.29 Complete model of a sheet-metal housing with the corresponding features structure and its mass characteristics

11.4 Modelling in NX

11.4.1 Selecting a Module for Modelling Sheet-Metal Products

Initiate a new model with the (*New*) command. Instead of a base model, select a sheet-metal product (*Sheet metal*) in the (*Model*) tab. This will lead to the (**.prt*) file type, allowing transitions between the base modules. An example is shown on Fig. 11.30, where the user begins modelling in the *Model* module and then switches to the sheet-metal module (*Application > Sheet Metal*) and saves the model in a file with a universal file format (e.g., *filename.prt*) (Fig. 11.31).

When modelling sheet-metal products, define the key parameters first in order to be able to logically define the specific conditions for specifying the cutting dimensions and end shapes separately for each specific shape. To enter the data, use the commands in the (*File > Utilities > Customer defaults*) menus. The basic parameters include the sheet-metal thickness, the bending parameters, etc. (Fig. 11.32). The shape-defining feature for the sheet-metal products already includes the deformation rules

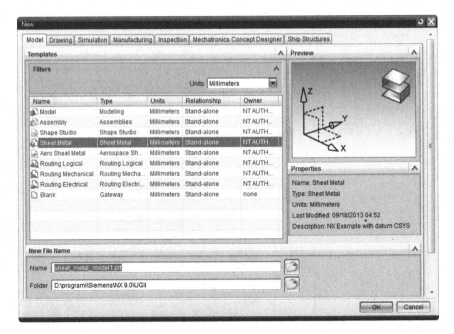

Fig. 11.30 Selecting the module for sheet-metal models (*Sheet Metal*) in the opening window

Fig. 11.31 Switching between different user modules (*Application* > *Sheet metal*)

and the tools, which define the shape of the details. If the manufacturing includes other special tools, they need to be included in the existing programme environment.

11.4.2 Creating the Base Shape

Each product is assigned its own base shape. Draw this base shape as a simple sketch of straight lines on the XC–ZC plane. Because all the key parameters of the sheet-metal material were already entered (Fig. 11.32), all the software programmes for this type of modelling allow the automatic formation of a 3D model with all of the corresponding roundings and deformations (Fig. 11.33).

The following procedure is used for setting the parameters in the shape details (Fig. 11.34).

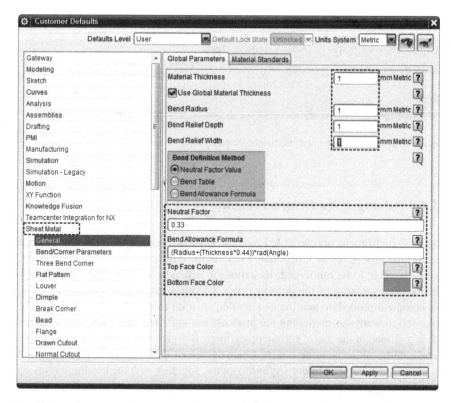

Fig. 11.32 Setting the parameters of sheet-metal material in the (*Customer Defaults*) menu

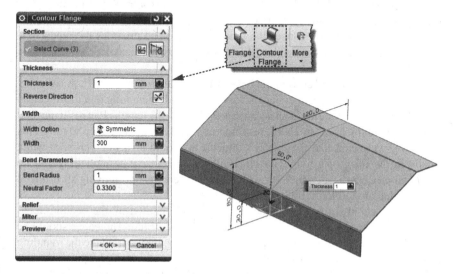

Fig. 11.33 Modelling a base shape with *simple lines* in the base sketch plane, including the main dimensions

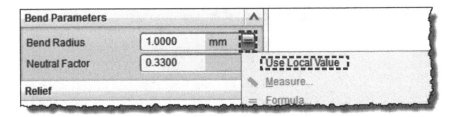

Fig. 11.34 Individual setting and resetting of the parameters for specific features, such as bending

11.4.3 Adding Edge Reinforcements

The edge reinforcements are added with the (*Flange)* feature. It is the name of the feature that represents as functionality details for reinforcement on a flange. First, define the edge to be reinforced. Its definition depends on the bending technology and the dimensioning of the deformations. Therefore, you need to know both the technological procedure and the tension-deformation model. In our case, it will be supposed that anyone modelling the products has sufficient previous knowledge or is being guided in the use of certain procedures. Below it will be necessary to verify the suitability of the bending radius, which can be either the same or larger than the default value in the (*Customer Defaults*) menu. Together with the bending radius on a particular edge, also enter the bending angle (e.g., 90°), the bending form (*Bend outside*) and the reference length (*Inside*), as shown in Fig. 11.35.

To connect corner flanges or to set or adjust them, use the (*Closed Corner*) feature (Figs. 11.36 and 11.37). On the Fig. 11.38 present the (*Trim body*) command.

Because it is a symmetrical model, mirroring is normally used for complex shapes. This significantly reduces the modelling time and provides the repeatability of the connected shapes. Execute the mirroring across the central XC–ZC plane (Fig. 11.39).

In order to close a model from the back, set a face at an angle of 90°, as shown in Fig. 11.38. Defining an accurate length for the face and orientating it in the direction of bending allows modelling at an accelerated pace. So, the length is set to 79 mm and the sheet-metal thickness to 1 mm, resulting in a total casing height of 80 mm. This case shows yet again that the model's dimension should always be close at hand, usually in the form of a rough sketch. It allows more accurate modelling and no additional procedures are required, such as adding lengths or removing sheet-metal elements that are too long (Fig. 11.40).

The products also include rounded edges, mainly caused by the bending technology. The shape is particularly pronounced on the outside. The edges should be rounded with the (*Break Corner*) feature, which is in principle similar to the rounding feature (*Edge Blend*) for the models (Fig. 11.41).

On the lower part of the casing create the reinforcing ribs with bent sheet metal using an angle of 90° and a length of 10 mm (Fig. 11.42). For this purpose, use

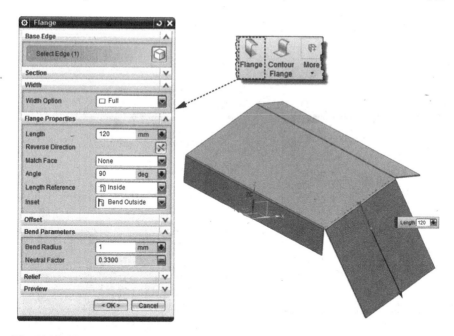

Fig. 11.35 Modelling the edge reinforcement or a flange with an angle of 90° using a reference inside length and the (*Flange*) feature

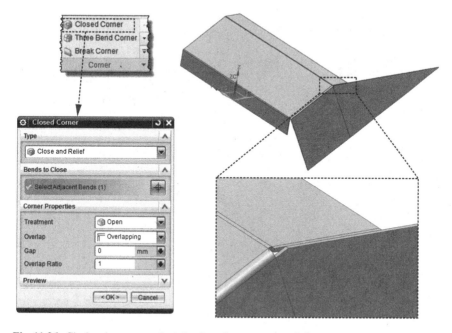

Fig. 11.36 Closing the connected reinforcing edges or corner reinforcements on a model with the (*Closed Corner*) command, and setting the overlap (*Overlap*)

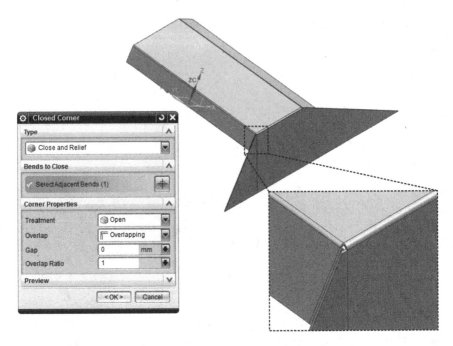

Fig. 11.37 Adjusting the edges in a corner with the (*Closed Corner*) feature

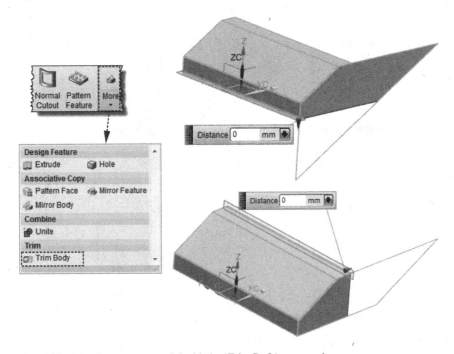

Fig. 11.38 Trimming excess material with the (*Trim Body*) command

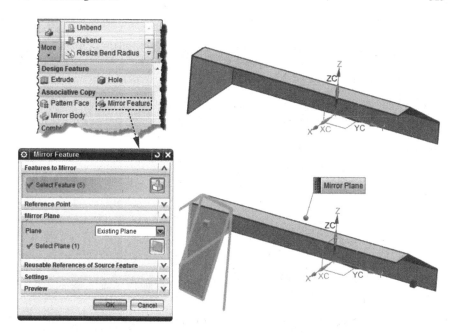

Fig. 11.39 Mirroring a side panel using the XC–ZC mirror plane

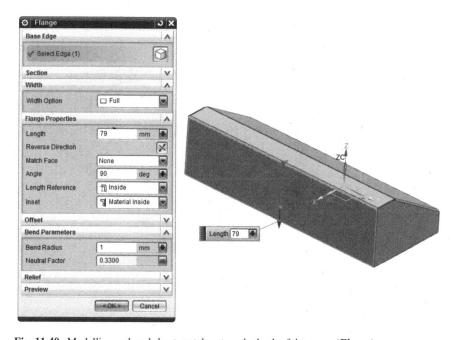

Fig. 11.40 Modelling a closed sheet-metal part on the back of the cover (*Flange*)

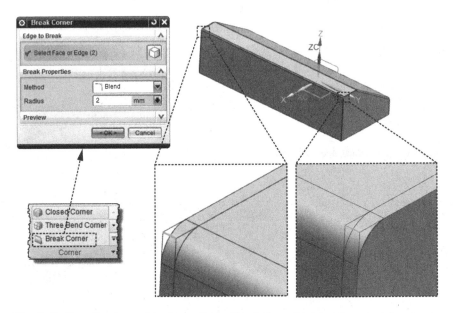

Fig. 11.41   Rounding the outside edges with the (***Break Corner***) command

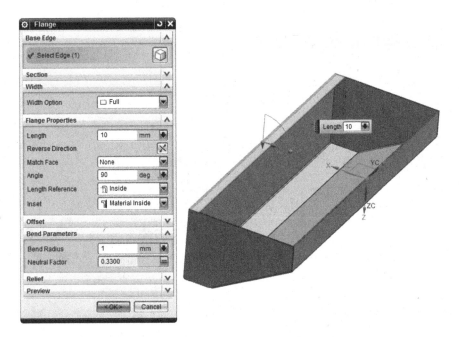

Fig. 11.42   Modelling the reinforcing edges on the lower part of the casing

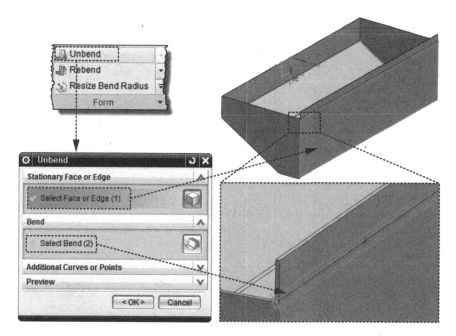

Fig. 11.43 Temporarily unbent flap (reinforced edge) with the (*Unbend*) feature

the (*Unbend*) feature for unbending a bent part. It allows the shape to be adjusted (Fig. 11.43).

When straight flaps overlap in the corners, they should be chamfered first. This is usually done at an angle of 45°, as it allows welding along the whole chamfer and thus provides significant reinforcement. For thinner sheet metals (less than 0.8 mm), these overlapping edges are not chamfered, but only bent by the same thickness on the other plane. This joint is then executed by spot welding.

The flat part can now be chamfered along the edges, as shown in Fig. 11.44, with the now familiar edge-blending (*Chamfer*) feature. Because the chamfer depth is not usually available, you can make use of direct measuring on the model. You can simply measure a distance on the model by clicking the button on the right of the window (*Distance*) and use it as default input data for the chamfer dimension. You can do all four reinforcements in this way, and then bend them again into the final position with the (*Rebend*) (Fig. 11.45) feature.

11.4.4 Completing the Shape

Before the model takes its final shape, complete the details that are important for both the shape and the functionality of the end product. Completing the shape is connected with the manufacturing technology, i.e., the technologicality. One of these

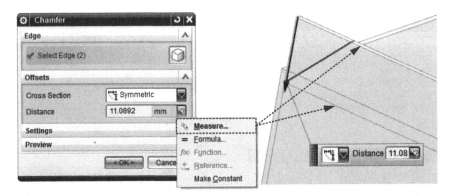

Fig. 11.44 Chamfer modelling on the corners (*Chamfer*)

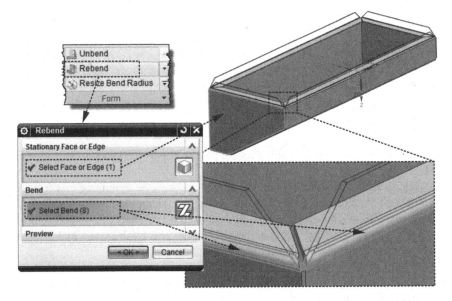

Fig. 11.45 Once all the edges (*flaps*) have been prepared for reinforcement and after chamfering, execute the bending of 90° for all such prepared flaps (*Rebend*)

final touches is important for including the attachment sockets (Fig. 11.46) and for other cut-outs (Figs. 11.47, 11.48 and 11.49) from the panels for inserting the parts into the casing. To present the individual steps and examples, figures from previous chapters are used.

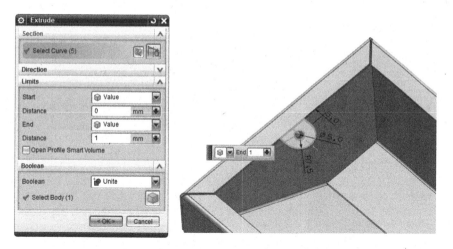

Fig. 11.46 Modelling of attachment sockets with the (*Extrude*) command

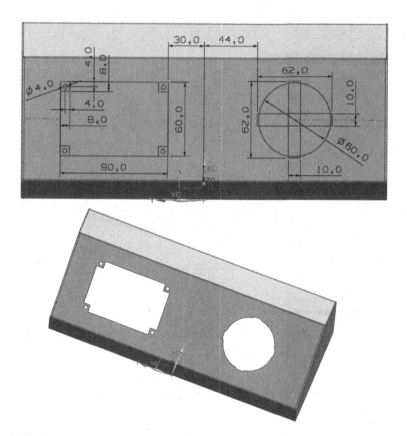

Fig. 11.47 Creating cut-outs on the casing's front panel with the (*Extrude*) command

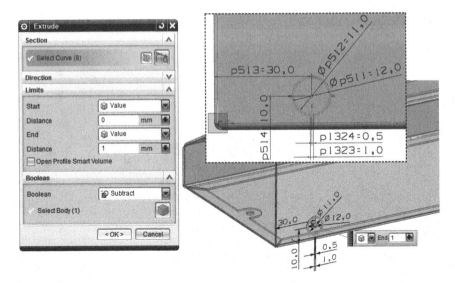

Fig. 11.48 Modelling cut-outs for the electrical feed-through connections on the back of the casing

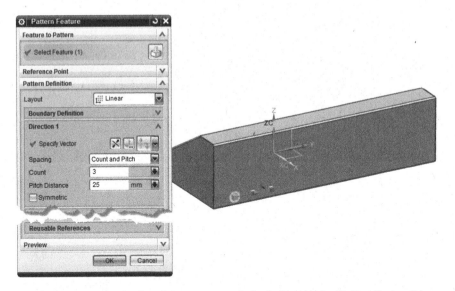

Fig. 11.49 Linear reproduction of the cut-outs on the back side with the (*Pattern Feature/Linear pattern*) feature

11.4.5 *Flat Surface of the Sheet-Metal Product's Final Shape*

The module for modelling sheet-metal products allows the flat sheet-metal surface to be defined. It significantly accelerates the preparation of the cutting documentation.

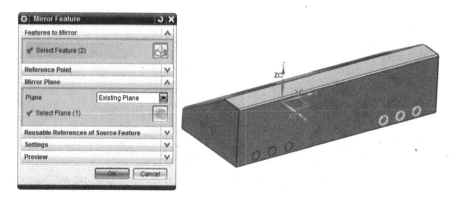

Fig. 11.50   Mirroring the cut-outs, modelled on the back of the housing, from right to left by means of the XC–ZC plane and using the (*Mirror Feature*) feature

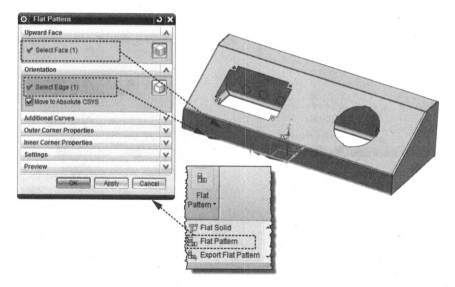

Fig. 11.51   The procedure for defining a flat sheet-metal surface with the (*Flat Pattern*) command

There are two options: to unfold the surfaces and to open the surfaces from a solid model (Fig. 11.50).

The first option is using the (*Flat Pattern*) feature (Fig. 11.51). It creates a 2D contour, ready for export to the DXF or DWG formats and puts it in the (*Part Navigator > Model View*) feature navigator under the default name (*FLAT-PATTERN#1*) (Fig. 11.52).

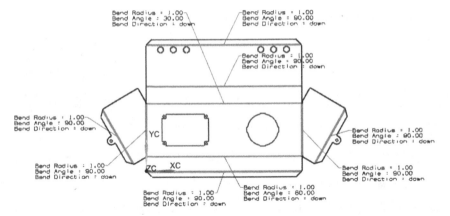

Fig. 11.52 The contour of a flat sheet-metal surface in the tree structure (***Part Navigator > Model View***)

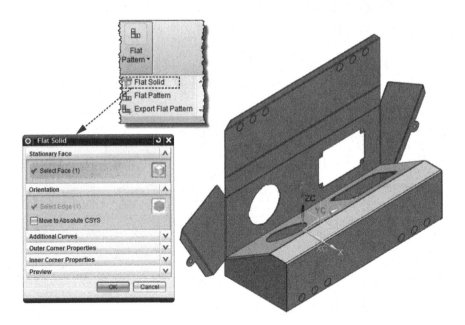

Fig. 11.53 Unfolding the sheet metal into a 3D flat body with the use of the (***Flat Solid***) feature

The other option for unfolding sheet metal is a method where the modeller actually unfolds the model into a 3D flat model (***Flat Solid***) (Figs. 11.53 and 11.54). This procedure creates a new body and the user can then hide or display any of the bodies (flat or bent).

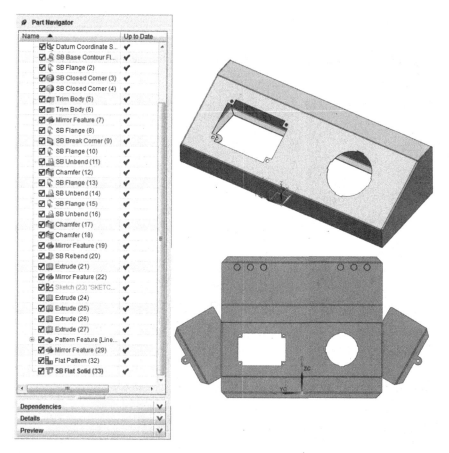

Fig. 11.54 The structure of the features tree and the end model in both the finished and flat forms

11.5 Examples

Figures 11.55, 11.56 and 11.57 present few examples for sheet metal bending.

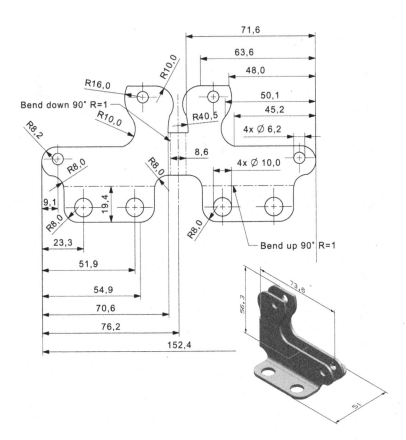

Fig. 11.55 Sheet metal housing support

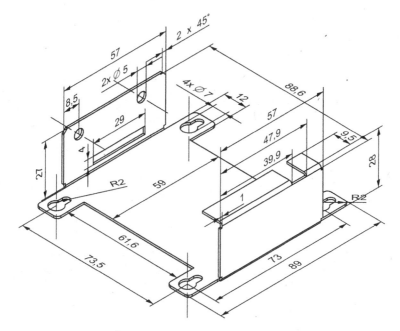

Fig. 11.56 Sheet metal housing support

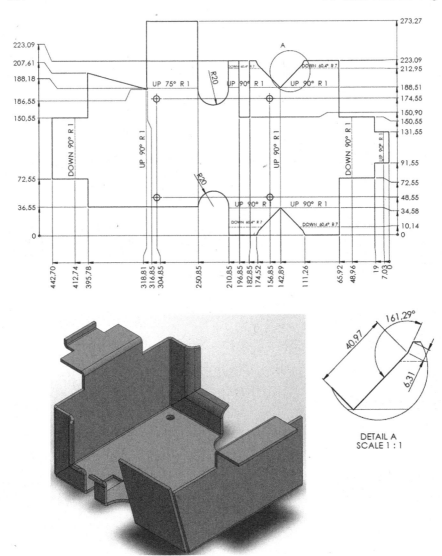

Fig. 11.57 Model of an unfolded housing

Chapter 12
Modelling Physical Models and Parameterization

Abstract The parameterization of individual parts is at its most important when using standard parts or parts from a range of products with the same main function, but changing loads and, indirectly, also dimensions. The methods are known through the use of tables or measurements. In each case, the procedures are presented, as well as the reasons why a particular parameter should be applied directly and why different values of the functions should be used. Various digitizing methods are presented, from traditional to modern laser-measuring instruments.

In this chapter we talk about different ways and approaches of dimensioning a physical model with a measuring instrument; we also talk about sketching the product and the transfer of data to a CAD model. We use this process for a mechanical part that does not have a CAD model. We also use it with existing products and devices that we wish to use in a new, complex component. To ensure the quality and the traceability we develop a new component virtually.

The chapter considers two types of actions. In the first part it is important for a constructor to physically dimension the part by him or herself. We present the simplest dimensioning approach and an approach that uses a laser. In any case, after the dimensioning the constructor needs to produce a sketch in 2D or 3D. This is left to the constructor to decide, but it is faster to sketch simple parts in isometrics and more complex parts in three views, so that all details can be seen. The sketching must aim to be a proportional presentation, which is presented in detail in Chap. 2. After a few attempts at mirroring the sketches the constructor can understand the proportionality and is competent to reproduce it. There are enough opportunities during the time of the study for a student to obtain this knowledge and to expand his or her talent to a satisfactory level. We can identify the ability of managing space by the sketching of simple models. After the sketching it is a good idea to model a rough shape of the product first, and later model all the details where the constructor needs to recognize the importance of the individual details. The sketching of a dimensioned part is crucial for an understanding of the product as a whole.

J. Duhovnik et al., *Space Modeling with SolidWorks and NX*,
DOI: 10.1007/978-3-319-03862-9_12, © Springer International Publishing Switzerland 2015

Fig. 12.1 Modelling a physical model of a screw M 20 × 120 mm, parameterization and formation of the configurations

In the second part we execute a parameterization of the model. First, it is necessary to define the parameters that comprehensively define the model (Fig. 12.1). The number of parameters is defined regarding the complexity of the model. Later, we tie all the dimensions of the model to the individual parameter using equations. By changing the parameters we form new shapes or so-called configurations of the individual product.

12.1 Measuring Physical Models

We choose a measuring approach regarding the complexity of the model, its size and the demands for the accuracy of defining the shape. To measure the characteristic dimensions of simple shapes of prismatic or axisymmetric models we commonly use measurements with jaws (meter, calliper, micrometer, etc.). More recently, lasers have been used to define more complex shapes. To use this process we need to define a multitude of points, through which we generate a surface and we obtain a virtual model (Fig. 12.2).

The dimensioning precision of the shape of an object depends on the future use of the object. If we wish to produce a CAD model of an already-existing part that will be implemented in another assembly, it is sufficient to produce a rough shape with the main and connection dimensions. In the case when we produce a replica of a certain physical model, a refinement with detailed shapes of a CAD model is necessary.

Later, an example of producing a CAD model for a screw M 20 × 120 mm is presented (Fig. 12.3b). The screw is a simple object that can be measured using a calliper (Fig. 12.3a). There are more types of calliper (common one, the one with a

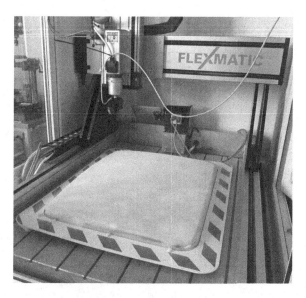

Fig. 12.2 Laser measuring

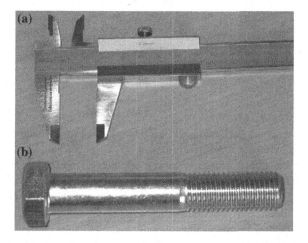

Fig. 12.3 Screw M 20 × 120 mm and Calliper (**a**) and Screw M 20 x 120 mm (**b**)

clock or a digital one) and various sizes of them. The most common ones are those in the range between 0 and 160 mm.

Callipers are general measurement instruments that make it possible to measure the outer and inner dimensions as well as the depth. The accuracy of the calliper is determined by the auxiliary scale on the movable part of the instrument—the vernier. There are decimal (1/10), one twenty (1/20) and one fifty (1/50) verniers (Fig. 12.4a).

Nowadays, we can also use a calliper with a digital screen (Fig. 12.4b) and a scale with an accuracy of 0.01 mm. This makes the accuracy of the measurement with

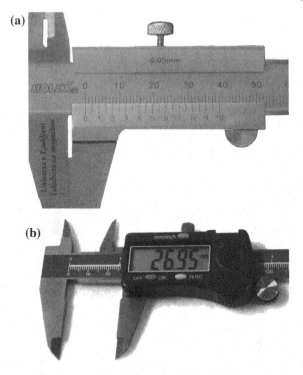

Fig. 12.4 One twenty (1/20) veriner (**a**) and calliper with a digital screen (**b**)

a decimal (measure a scale of ten segments) $1/10 = 0.1$ mm, with the one twenty (measure scale of twenty segments) $1/20 = 0.05$ mm and with the one fifty (measure scale of fifty segments) $1/50 = 0.02$ mm.

If we need even greater accuracy, we use micrometer, which has an accuracy of 10 microns (0.01 mm) or even 1 micron (0.001 mm) (Fig. 12.5). The idea is that for each type of measurement a different instrument is needed. There are outside, inside (Fig. 12.5b) and depth (Fig. 12.5c) micrometers.

Free-hand sketching is one of the elements that have an important part in a presentation and in an understanding of individual shapes. In the recent digital era this segment of activities in the techniques and engineering is often left out. It is also important for students to learn this kind of presentation.

Free-hand sketches need to present the shape in real proportions. It has to be obvious from the sketch wether the sketch is in an auxiliary geometry (thin line) or if it is a final shape (thick line).

2D sketches usually present various views and cross-sections and they have the dimensions included (Fig. 12.6). 3D sketches usually present the shape of the object in the isometric projection (Fig. 12.7).

After the measurement of the model we prepare a sketch that clearly presents the part we are working on.

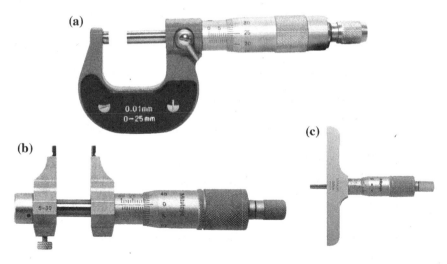

Fig. 12.5 Micrometer: outside (**a**), inside (**b**) and depth (**c**)

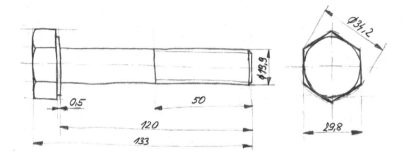

Fig. 12.6 2D sketch of a screw M 20 × 120 mm

12.2 Base Model

In the manufacturing process of a certain product we need to be aware of its use (the function) and its shape (the manufacturing technology, the ergonomics, the positioning in the space). In this way we produce products for a limited environment of use.

The larger capacity of the product and its capability for use with different energy sources or different users or a simple enlargement capability in regard to the user environment, quickly requires an overview of the possible new types of usage. These new demands allow us to check whether we are conceiving a new product or if the already known origin is widening because of a larger capacity or complexity. If a greater capability is possible with a simple enlargement of the product, it means the product is already close to perfection in terms of its function. This product can

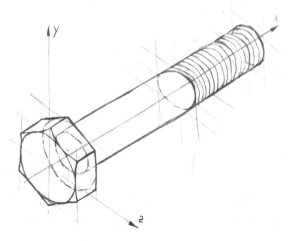

Fig. 12.7 3D sketch of a screw M 20 × 120 mm in the isometric

be developed in a modular design. It depends on the concept on which the product was constructed, i.e., if the modular design is by function or by geometry. A modular design is not only a construction from modules, but it is also the assembly by function or dimensions, as well as by shape.

The connection between the adjustment of dimensions and the capability of individual products is known. For this matter we use the arithmetic row to determine the steps between the individual dimensions for shapes. For functions we are aware of adjusting the capability with the geometric row. What kind of progress we use for a completely new product we can decide on our own.

For this reason the constructor needs to have the ability to construct files that provide a rapid logical adjustment of the dimensions (shape) and capabilities (function) in the selection of individual shapes and functions. In this chapter we first wish to present a measurement recording of the objects for the preparation of a digitalization and subsequently we present the principle to the parameterization of the considered parts.

The process of the parameterizational modelling of the product is as follows:

- In the first step we need to model the starting model that will be used as a template for similar products.
- Subsequently, we need to define the rough geometric parameters that define the product (e.g., for a screw we need to define the diameter of the body).
- We determine the number of parameters with regard to the complexity of the model. We link the specific detailed shapes with the basic parameters using equations. The user decides about the importance of the specific details. For example, for the thread of a screw we ask ourselves whether a detailed shape is needed or if a symbolic mark is enough.

Fig. 12.8 Parametric definition of screw

- We define the chosen parameters in the so-called generator of parts and we determine the locations for the file's generation with models in a computer.
- We form the new shapes or configurations of the individual product by adjusting the parameters.

12.2.1 Screw Parametrization

Below, we will first look at model parameterization. To do this, you first need to define the basic parameters (dimensions) that affect the shape and function of each product. It should be noted that parameters are often executed directly from equations, describing a function of a particular process. For standardized elements, they are executed from tabular dimensions or dimensions, defined by ratios.

On the basis of the main parameters, which you can set by yourself for new products, and by means of mathematical connections and equations, we will define other dimensions, required to describe a model. Creating the details, this will later allow us to link up their parameters (dimensions) and the basic parameters.

First, you should define the basic parameters, describing our object (Fig. 12.8). In the case of the screw, they are as follows:

- **screw diameter** ($d = 20$ mm)
- **screw length** ($l = 120$ mm)

The executed dimensions (Tables 12.1 and 12.2), defined by a ratio, are as follows:

- **thread length** ($b = 50$ mm, $b = (2.0; 2.5; 3.0) \times d$)
- **thread lead** ($P = 2.5$ mm, defined by standard, relative to thread size; it is roughly $P = 0.125 \times d$, thread dimensions are then rounded off to two decimal places if the measurement is in millimeters
- **wrench opening** ($s = 30$ mm, $s = 1.5 \times d$ based on the diameter of a hexagon head, encircled by hexagon's circle and defined as $D_g = (1.8$ to $2.0) \times d$; for IMBUS screws or screws with a cylindrical head, the screw head diameter is around $D_{gValj} = 1.6 \times d$)
- **screw head height** ($k = 13$ mm, $h = 0.7 \times d$, nut head $m = 0.8 \times d$)

Table 12.1 Table of main parameters of a screw line (dimensions in mm)

	M 20 × 120	M 16 × 60	M 12 × 90	M 10 × 70	M 10 × 30
Cylinder diameter—d	20	16	12	10	10
Cylinder length—l	120	60	90	70	30
Thread length—b	50	40	30	30	30
Thread lead—P	2.5	2	1.75	1.5	1.5
Spanner opening—s	30	24	19	17	17
Head height—k	13	10	7.5	6.3	6.3

Table 12.2 Metric threads data

Thread	Pitch P	Major D/d mm	Pitch Dia D_2/d_2 mm	Minor D_1/d_1 mm
M3 × 0.5	0.5	3.000	2.675	2.459
M4 × 0.7	0.7	4.000	3.545	3.242
M5 × 0.8	0.8	5.000	4.48	4.134
M6 × 1.0	1	6.000	5.35	4.917
M8 × 1.25	1.25	8.000	7.188	6.647
M10 × 1.5	1.5	10.000	9.026	8.376
M12 × 1.75	1.75	12.000	10.863	10.106
M16 × 2.0	2	16.000	14.701	13.835
M20 × 2.5	2.5	20.000	18.376	17.294
M24 × 3.0	3	24.000	22.051	20.752
M30 × 3.5	3.5	30.000	27.727	26.211
M36 × 4.0	4	36.000	33.402	31.67

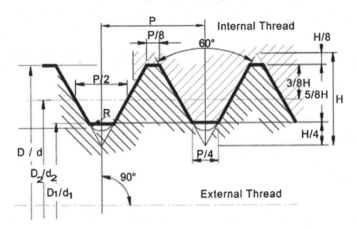

Fig. 12.9 Metric threads data

As a general rule, ratios are important to quickly define the main dimensions. More advanced modelers have the shape dimensions that depend on the main parameter presented in Table 12.1.

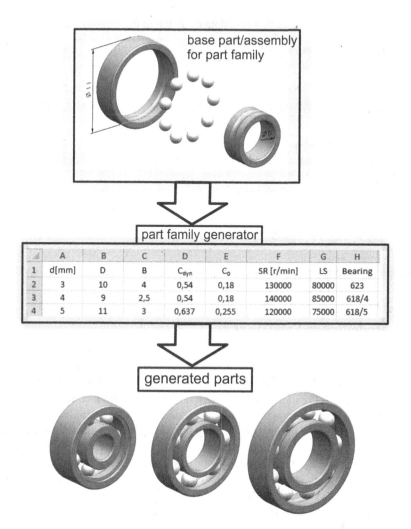

Fig. 12.10 The concept of the formation of similar products using the parts generator

In the case of a set of geometrically similar products with similar dimensional relations, i.e., the shape of the model does not change, you should strive to prepare such products into a line—configuration. The term "line" denotes all products, composed of the same features but of different dimensions. In addition to differences in dimensions, configurations also have some suppressed features. You can read dimensions and features from specially adapted tables.

In order to generate new configurations, activate the configuration manager (*Configuration Manager*). It is important to be able to apply different configurations to the created model. They are edited or entered with the same command (*Configuration Manager*).

This configuration allows adding new data or updating the existing data.

- Thread data at different screw sizes (Fig. 12.9 and Table 12.2)
- Create a spiral which will be used as the curve of extrusion.

  - the basic circle is identical to the diameter of the screw cylinder
  - thread lead (see Fig. 12.9)
  - thread length

- Draw the thread profile
  Select the data from the Table 12.2 and relate them by means of equations (parametricity).

12.2.2 Bearing Parameterization

Principles of bearing parameterization is shown on Fig. 12.10.

12.3 Modelling in SolidWorks

The modelling of existing products in SolidWorks is identical to the processes that were presented in previous chapters. We select the appropriate approach with regard to the specificity of the individual object. When we deal with a certain product we firstly measure it and later we produce its sketch. While sketching we control the comparable features for the space modelling that should ensure the accuracy and the reality of the digital presentation of the product to a greater extent. In the case of modelling a screw we produce a rough shape using an extrusion. For a detailed shape we improve the rough sketch with a revolving, a curve sweep and other additional features (a fillet, a chamfering, etc.).

For the parameterization we firstly define the main dimensions, i.e., the parameters on the model. We later tie the rest of the dimensions to the existing parameters using links and equations. By changing these parameters we produce various configurations in SolidWorks or in Microsoft Excel.

12.3.1 Rough Shape

At the beginning of the modelling we first introduce a shape as a rough structure of simple geometrical shapes. We have to ensure that all the main measures are presented and that the connecting details are displayed clearly and accurately. In our case the main shapes are the cylinder that represents the body of the screw and the hexagonal prism that presents the head of the screw.

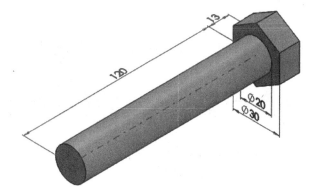

Fig. 12.11 CAD model of a rough shape of the screw M 20 × 120 mm

To model the body of the screw we draw a circle with a diameter of 20 mm in a basic plane (front view) and we extrude it into a space for a value of 120 mm. In a similar way we produce the head. We draw a hexagon with an inner diameter of 30 mm on the same plane and we extrude it into space with a value of 13 mm (Fig. 12.11).

12.3.2 Detailed Shape

In the detailed shape of the screw we include the chamfering at the end of the body, the fillet between the body and the head, the chamfering on the head, the modelling of the thread, etc.

At the end of the body of the screw we produce chamfering (***Chamfer***) in regard to the nominal diameter. For screws we use chamfering with an angle of 45° and a length that is the same as the pitch of the thread. For that reason we always need to check the height of the pitch of the thread on the part of the thread when we draw the chamfering of the screw. We also need to mark this pitch. This is important because the height of the pitch also determines the depth of the thread or the diameter of the core of the screw (the loading cross-section). In our case we have the screw M 20 with the pitch of the thread P = 2.5 mm, so we implement the chamfering with 2.5 mm/45°.

We produce the transition between the body and the head of the screw with a ***Fillet***. We measure the size of the fillet on the model. In this case the screw is measured and fillet has a size of 1 mm.

The head of the screw is also chamfered from both sides. The easiest way to produce this shape is by revolving (***Cut Extrude***). We draw a sketch (we define the parameters for the chamfering in relation to the original size of the head or to the measurement) and we deduced this sketch from the head of the screw (Fig. 12.12).

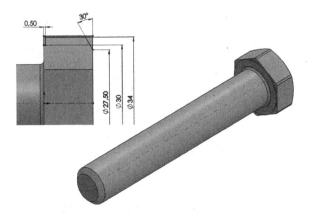

Fig. 12.12 Chamfering and fillet on the model of the screw M 20 × 120 mm

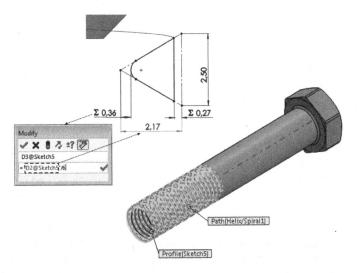

Fig. 12.13 Modelling the thread M 20 and the formation of the profile of the thread

We usually do not model threads in a geometric shape but we introduce them as a cosmetic thread (***Insert > Annotations > Cosmetic Thread***). However, in our case we model the thread. The process of producing the thread is similar to the process of producing the spring. The only difference is that here we have a function of the subtraction of the material (***Swept Cut***). We firstly form a spiral (height = 50 mm, pitch = 2.5 mm) that we will use as a curve sweep. The basic circle is equal to the diameter of the body of the screw M 20. We draw a sketch of the profile of the thread in the rectangular plane (top view). We then select data from Fig. 12.13.

We still need to produce an expiry of the thread. We execute a ***Revolved Cut***, where we consider the profile of the thread as a sketch (***Convert Entities***). Figure 12.13

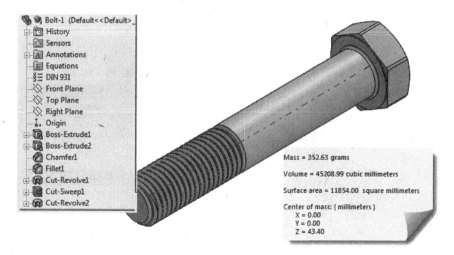

Fig. 12.14 Model of a detailed shape of the screw M 20 × 120 mm with its structure of features and mass characteristics

shows a detailed model of the screw M 20 × 120 mm with its structure of features and mass characteristics (Fig. 12.14).

12.3.3 Parameterization

Next, we will consider the parameterization of the model. For this reason we define the basic parameters (dimensions) that have the main influence on the shape and the function of the individual product (Fig. 12.15).

On the basis of the main parameters that can also be defined by ourselves when dealing with new products, we define the rest of the dimensions that are needed for the inventory of the model. We use mathematical connections or equations. This enables us to link the parameters (dimensions) to the basic parameters later on during the creation of details.

We link the chamfering on the body of the screw (2.5 mm/45°) with the pitch of the thread so that the length of the chamfering equals the pitch of the thread ($D1@Chamfer1 = P@Helix/Spiral1$). We define the fillet between the body and the head of the screw (R = 1 mm) as 1/20 of the diameter of the body ($D1@Fillet1 = 1/20*d@Sketch1$) (Fig. 12.16).

We parameterize the chamfering on the head of the screw in the sketch. We link the dimensions that determine the sketch to the basic parameters and we logically form the connective equations. Figure 12.17 shows the equations that we use for the parameterization of the chamfering on the head of the screw.

The parameterization of the thread was already partially produced in the sketch where we lined individual dimensions to the height of the thread. We still need to

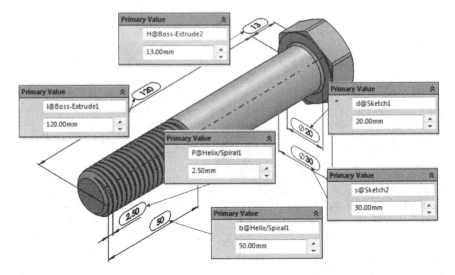

Fig. 12.15 Parameterization of a model of the screw

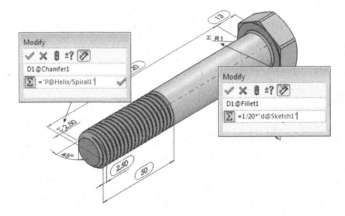

Fig. 12.16 Parameterization of the chamfering and the fillet on the model of a screw

link the pitch of the thread in the sketch of the profile with the parameter of the pitch of the thread in the forming of the spiral (Fig. 12.18).

Figure 12.19 shows the equations that link the basic parameters to the rest of the dimensions that describe the model of the screw.

12.3.4 Generation of a Configuration

To generate new configurations we activate the **Configuration Manager**. We form the individual configuration by selecting the model and adding the new configuration

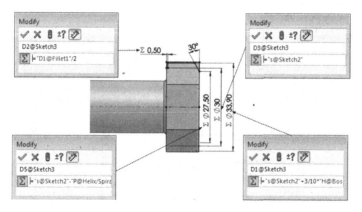

Fig. 12.17 Parameterization of the chamfering on the head of the screw

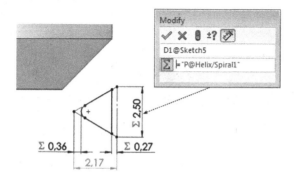

Fig. 12.18 The link of the pitch of the thread in the sketch of the thread to the pitch of the spiral

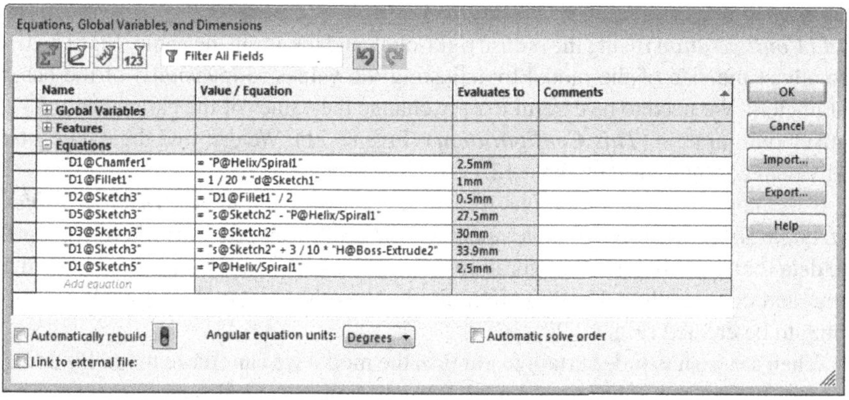

Name	Value / Equation	Evaluates to	Comments	
⊞ Global Variables				
⊞ Features				
⊟ Equations				
"D1@Chamfer1"	= "P@Helix/Spiral1"	2.5mm		
"D1@Fillet1"	= 1 / 20 * "d@Sketch1"	1mm		
"D2@Sketch3"	= "D1@Fillet1" / 2	0.5mm		
"D5@Sketch3"	= "s@Sketch2" - "P@Helix/Spiral1"	27.5mm		
"D3@Sketch3"	= "s@Sketch2"	30mm		
"D1@Sketch3"	= "s@Sketch2" + 3 / 10 * "H@Boss-Extrude2"	33.9mm		
"D1@Sketch5"	= "P@Helix/Spiral1"	2.5mm		
Add equation				

Fig. 12.19 Equations for the parameterization of the model of the screw

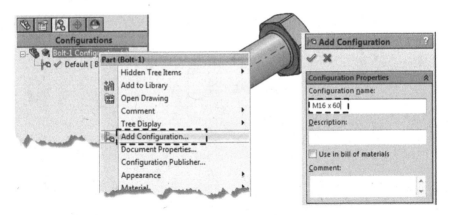

Fig. 12.20   Generation of the individual configurations

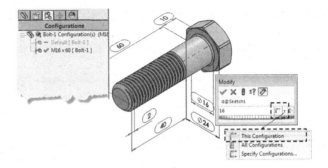

Fig. 12.21   Adjusting the values of parameters directly on the model

(***Add Configuration***) using the mouse right-click and inputting the name (Fig. 12.20). We adjust the size of the model by adjusting the values (dimensions) of the basic parameters. We need to be careful to only change the values of the parameters of the active configuration (***This Configuration***) (Fig. 12.21). We can find the data of the individual configuration in Table 12.1.

If we have more configurations we can use construction tables (***Insert > Tables > Design Table***). The Excel table opens in the SolidWork program where we input the data for the individual configuration in an individual line. When we complete the input and confirm the table, the program alerts us that inputted configurations are going to be created (Fig. 12.22).

When we wish to hide certain features in the model we can freeze them. By doing this we obtain a new shape for the model. In addition, we can also produce a *Derived Configuration* based on a certain configuration (Fig. 12.23).

Figure 12.24 shows the shapes of screws that are configured using the configuration tree. From the ***Default*** configuration we first produce the basic configurations. Later, we create the derived configurations from them in regard to the display mode.

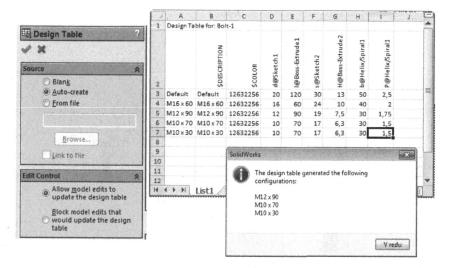

Fig. 12.22 Generation of the configuration using a Microsoft Excel table

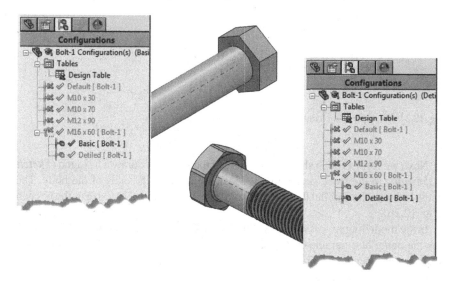

Fig. 12.23 Derived configuration of the screw M 16 × 60 mm

12.4 Modelling in NX

12.4.1 Creation of a Rough Shape of the Model

Before each modelling we first need to form a rough shape of the product we wish to model. We start the modelling with the main parameter. In our case this is the body with a diameter of $\phi 20$ mm and its length—the body of the screw (Fig. 12.25).

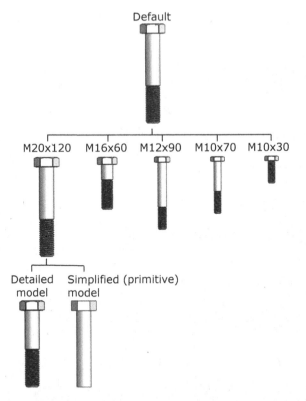

Fig. 12.24 Example of the configurations on the model of the screw

Next, we form a rough shape of the head of the screw. We draw a hexagon (***Sketch > Polygon***) in the basic sketch that is located on one of the ends of the body. The width between the parallel flanges is 30 mm, and this represents the size of the wrench (Fig. 12.26).

In the modelling process we notice that each dimension that we form in the sketch has the name of a parameter. It starts with a small letter ***p***, which is followed by a sequence number. Each parameter we define is automatically inputted in a special list that can be accessed in the main menu (***Tools > Expressions***) (Fig. 12.27). We can rename these parameters or we can use them as the variables for optional functions.

In the open window for the determination of expressions or parameters we can use the default names for them or we can rename them. We put a desired name next to each parameter in the row marked ***Name***. As a space between the words we use the underscore character. We can input the equation where the parameter is a variable in the row named ***Formula***. In this way we add a characteristic parameterization to each parameter. The calculated value is displayed in the row ***Value***. In the next steps we define the parameters that were not yet used. We wish to have them defined in the parameterization in details (Fig. 12.28).

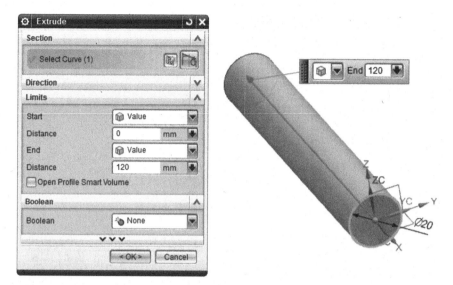

Fig. 12.25 Modelling of the body of the screw with a diameter $\phi20$ mm and a length for the body of 120 mm

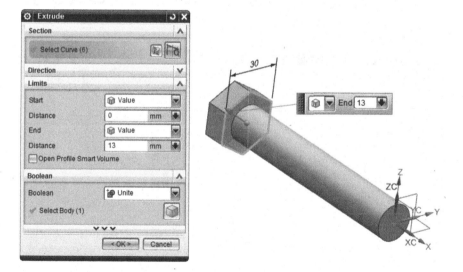

Fig. 12.26 Modelling a rough shape of the head of the screw

12.4.2 Formation of a Detailed Shape of the Model

We first produce detailed shapes on the head of the screw and later on the body. These details are linked to the main parameters with simple equations. We previously named these parameters (Fig. 12.28):

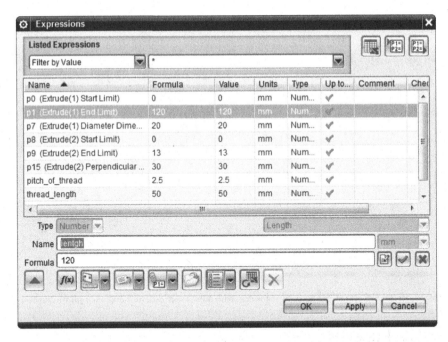

Fig. 12.27 List of the defined parameters (*Tools > Expressions*)

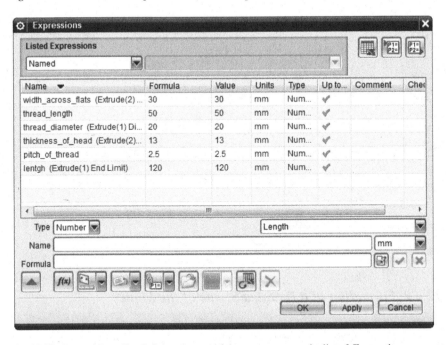

Fig. 12.28 Renaming and/or defining the rest of the parameters on the list of *Expressions*

Fig. 12.29 Definition of the
dimensions of detailed shapes

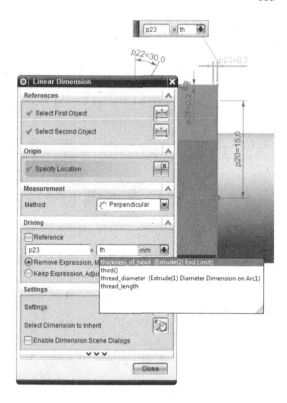

- *Width across flats* = 30 mm
- *Thread length* = 50 mm
- *Thread diameter* = 20 mm
- *Thickness of head* = 13 mm
- *Pitch of thread* = 2.5 mm
- *Length* = 120 mm

To define the individual feature we use a pre-selected parameter, which we pre-viously define (example: screw diameter). We can also input a function where the name of the parameter is a variable (Fig. 12.29).

Now that we have defined all the essential parameters we start modelling a rotation contour by removing the material on the head of the screw (Fig. 12.30).

We produce the modelling of the head of the screw by revolving (***Revolve***) (Fig. 12.31).

On the other side of the body of the screw we perform a chamfering where we use the defined parameter ***Pitch of thread*** for the depth at an angle of 30° (Fig. 12.32).

To model a thread on the body of the screw we perform a curve sweep from the basic sketch of the thread's cross-section and we implement it on a certain flank of the thread using the command (***Helix***). As an origin of the flank we use the free end of the body. For the dimension of the pitch of the thread (***Pitch***) we choose a

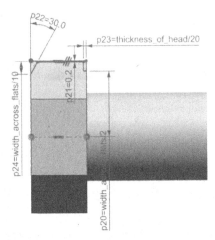

Fig. 12.30 Formation of a detailed shape of the head using the parameterization dimensions that correlate to the basic parameters

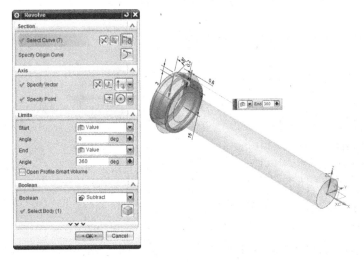

Fig. 12.31 Modelling a chamfering of the head of the screw by rotation of the sketch around the axis of the body

pre-selected parameter **Pitch of thread**. To determine the length of the thread we use the command **Limits** to select the origin value 0. For the final value we use the parameter **Length**. We choose the right-hand turn direction of the flank of the thread (Fig. 12.33).

To model the basic sketch that represents the profile of the thread we link the dimensions of the profile of the thread to the basic parameters. This process is the same process that is determined when defining the dimensions for a standard thread.

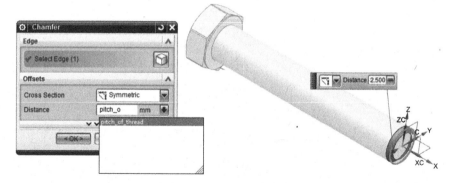

Fig. 12.32 Modelling of a chamfering at the end of the body of the screw

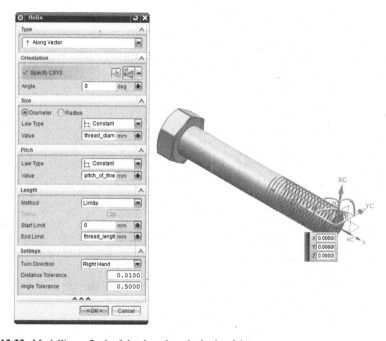

Fig. 12.33 Modelling a flank of the thread on the body of the screw

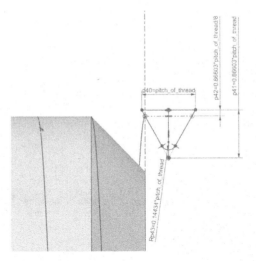

Fig. 12.34 Modelling the basic sketch of the profile of the thread

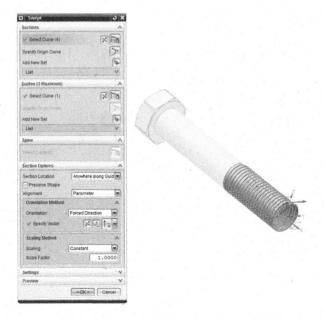

Fig. 12.35 Curve sweep of the basic sketch that represents the profile of the thread on the spiral and the final result of modelling the thread on the screw

The shape of the thread is shown in Fig. 12.34. We perform a curve sweep into space on the main curve of the spiral that we formed previously (Fig. 12.35). Thread is finalized with revolve feature as shown on (Fig. 12.36).

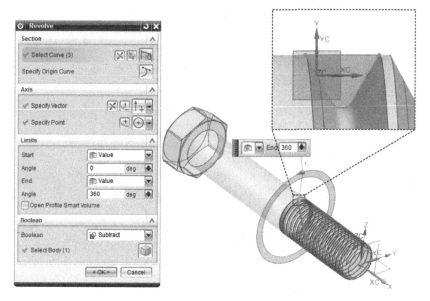

Fig. 12.36 We perform a rotation contour of the cross-section of the thread and we remove the material to obtain the final shape of the thread

12.4.3 Formation of the Generator of Parts

It is a special challenge to set the generator for families of products (e.g., screws) that allows us to choose between different parameters and to change them. With this model of the generator we can form 3D models of similar products and we can save them in the selected folder.

We activate the settings for the generator of parts (*Tools > Part Families*) in the main menu and we systematically mark those parameters that we wish to use in the modelling process. We choose from the window named *Available Columns*. We mark the parameters using the button *Actions* and we move the selected parameters into the window named *Chosen Columns*. We use the command *Create Spreadsheet* to finish (Fig. 12.37). With this command we create a table from which we can choose the dimensions. It also allows us to create parts that have the selected generic model. On the bottom of the window *Part Families* (Fig. 12.38) we select the folder where we wish to save the generic parts.

We select the tab **Add-ins** from the window with the table that automatically opens (Fig. 12.36). We obtain a list where the first row shows the number of the part and the second row displays the name that will be used for the generic part (Fig. 12.39). The rest of the rows we have already defined in the generator (*Part Families*), where we have selected the desirable parameters. By selecting the entire line with data and activating the command *Part family > Create part* we create a

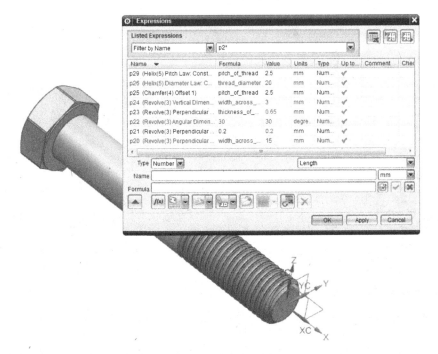

Fig. 12.37 The final basic shape and the detailed shape of the screw

generic part. Now we can check the generic model of the screw in the selected folder *Family save Directory* that we defined in the generator of parts.

If we wish to turn on/off certain geometric shapes in the generator of parts we execute the following process.

Firstly, we select a group of features that form a certain shape (for example, a thread) and we activate the command *Feature Group* using a mouse right-click. Now we have all the features that form our shape selected and we unify them into the group of features, to which we define a name. We activate it in the generator (*Part Families*) in a similar way to how we activate other parameters (Fig. 12.40).

We can also use this function for complex shapes.

We transfer the selected group of features of the thread into the window of the selected parameters (*Chosen Columns*) and we activate the table (*Spreadsheet*). This allows us to optionally turn on/off the shape of the thread in our case (Fig. 12.41). At the end we have parameterization structure as is shown on Fig. 12.24.

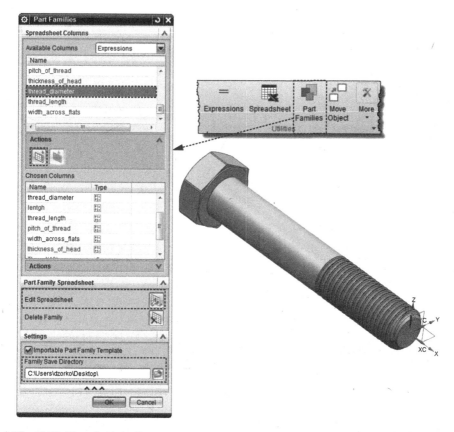

Fig. 12.38 The selection of important parameters in the parts generator

	DB_PART_NO	OS_PART_NAME	thread_diameter	lentgh	thread_length	pitch_of_thread	width_across_flats	thickness_of_head
1								
2	M20x120	M20x120	20	120	50	2,5	30	13
3	M16x60	M16x60	16	60	40	2	24	10
4	M12x90	M12x90	12	90	30	1,75	19	7,5
5	M10x70	M10x70	10	70	30	1,5	17	6,3
6	M10x30	M10x30	10	30	30	1,5	17	6,3
7								
8								

Fig. 12.39 Table for selecting the dimensions for the formation of the individual parts

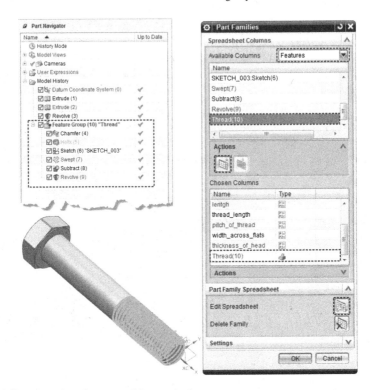

Fig. 12.40 Integration of a group of features in the generator

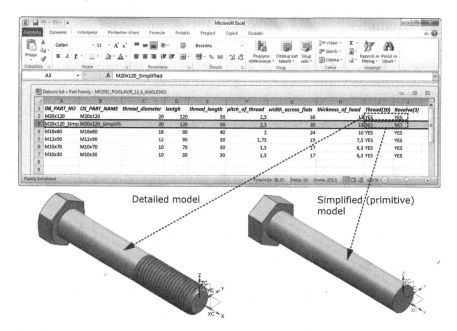

Fig. 12.41 Turning on/off a thread as a shape combined into the group of features

12.5 Examples

Figures 12.42, 12.43 and 12.44 present few examples for parameterization.

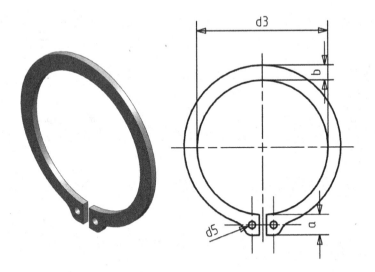

d3	a	d5	b	t	[mm]
18,5	4	2	2,6	1,2	
20,5	4,2	2	2,8	1,2	
22,2	4,4	2	3	1,2	

Fig. 12.42 Produce the parametric model for the mechanical element—a snap

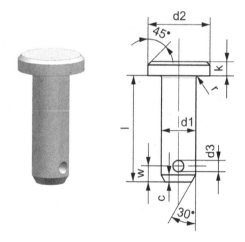

d1	d2	d3	l	k	r	w	c	[mm]
3	5	0,8	6	1	0,6	1,6	1	
4	6	1	8	1,6	0,6	2,2	1	
5	8	1,2	10	2	0,6	2,9	2	

Fig. 12.43 Produce the parameter model for the bolt

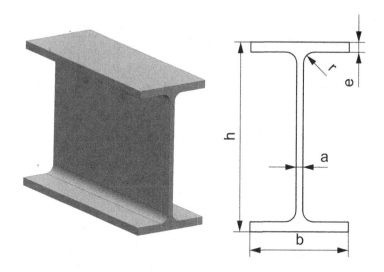

h	b	a	e	r	[mm]
80	46	3,8	5,2	5	
100	55	4,1	5,7	7	
120	64	4,4	6,3	7	

Fig. 12.44 Produce an example for the HEA profile (ISO standard)

Chapter 13
Assemblies

Abstract The last two chapters are dedicated to the technologies of creating working documentation. This chapter presents the bottom-up and top-down techniques. A theoretical model for an understanding, and the connection with the abstraction of the function, which should be performed individually for each product or assembly, is presented. A model of a shaft with bearings is presented again, followed by a description of how to cover it and install it in a casing. This shaft example proves how both methods logically complement, rather than exclude, each other.

13.1 Technology

The modelers that allow the space modelling of a product during its entire research and development cycle currently still lack the capabilities to comprehensively create the technical documentation. It consists of technical reports, studies, material specifications, setting manufacturing processes, and the corresponding documentation for shipping, transport, distribution, after-sales service, servicing, and product disposal after use. The integrating information during the entire life-cycle means presenting the design, i.e., from a rough model to a detailed model, structural and technological plans, to the photographs of models or real products themselves. Some modellers' modules provide high-quality presentations, while comprehensive presentations require the use of presentation standards and up-to-date presentation technologies.

The integration of presentations in the R&D process and the current state of existing technologies is well known and will not be dealt with further. Instead, the situation will be recapitulated and logically applied to quality and direct use. Within the modelling basics, it is not our attention to develop or present the sensibility and concepts of links for the purpose of creating technical documentation. However, we would like to show how to logically use a digital model of a product from its beginning to the detailed plans, important for the creation of a product and enabling

J. Duhovnik et al., *Space Modeling with SolidWorks and NX*, 367
DOI: 10.1007/978-3-319-03862-9_13, © Springer International Publishing Switzerland 2015

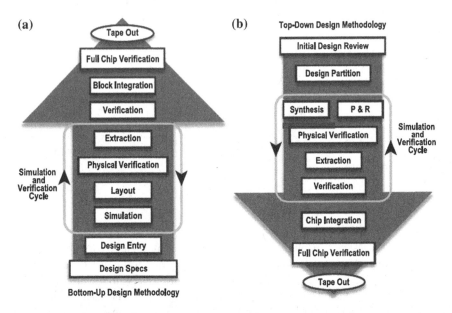

Fig. 13.1 Design process methods: bottom-up (**a**) and top-down (**b**). *Source*: Verification academy cookbook

the traceability of all the key parameters in order to increase and, most of all, to maintain the quality of manufacturing and servicing.

In the previous chapter, two methods for devising a product were indicated: (1) bottom-up and (2) top-down. Each of them applies to different procedures of research, development or product engineering/designing (Fig. 13.1).

The bottom-up method is typical and mostly used for studying the technological processes of new states or new processes. In such cases we in principle deal with both new functions for a specific use and new shapes. Because new functions should be given their shapes as soon as possible, it makes the designing and analysis of the states in space all the more important. Simplifications, expressed by transformations into planar space, are nothing more than simplifications, and do not represent any significant and fully sensible influence upon the space. For this reason, these cases in particular call for an analysis in space with all of the important support analyses. The building therefore takes place from an element to a part, a sub-assembly and to an assembly (Fig. 13.1a). This division is important in order to master the manufacturing, assembly, distribution and completing of objects on-site, as well as to understand the establishing of the structure of managing technical documentation in general. It is a well-known fact that manufacturers of door locks cannot understand the process of building an object. On the other hand, those who are familiar with large facilities cannot recognize the details and concepts of producing the locks for the doors of those same buildings, although they can appear as owners in separate production systems or separate economic subjects, comparable with their competition.

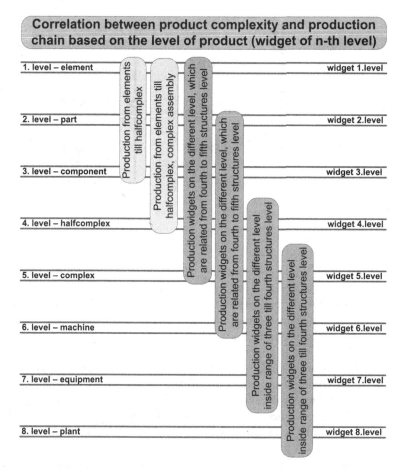

Fig. 13.2 Product structure in the global products concept

The top-down method represents an option for product development with pre-set functions and tailor-made works elements (Morphological matrix, MFF matrix or TRIZ). Each working principle, recognized in advance, can be recognized in detail or as a standard element (bolts, bearings, electric motors, small gearings, etc.). The method is very useful for standard parts and it allows the relations to be defined in advance. Relations are defined geometrically, e.g., parallel, perpendicular, centric, tight fitting, transition fitting, etc. Such defined relations are then defined in a modelled assembly, which is then fitted with standard elements that have defined relations. When dealing with inventive engineering that generally integrates and uses in the composition the already-known elements with known working principles, the top-down method is excellent (Fig. 13.1b). It goes well with the morphological matrix. The next level is the matrix of function and functionality (MFF). It allows structural building with a constant evaluation of the selected working principles.

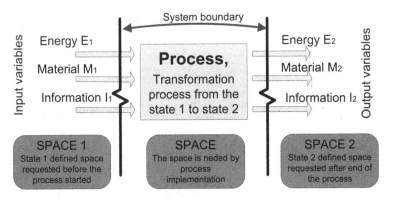

Fig. 13.3 Function defined by energy, material, information and form of the product, defined by space transformation through the process

In each case, you need to prepare for each assembly an adequate structure and integration between the levels of building blocks. In production and manufacturing systems there are specific groups of levels that can be integrated within a single process. With more than five levels of building blocks in one finished unit it is not possible to expect manufacturing integration (Fig. 13.2).

In the R&D process the goal is always to define a product in the best possible way, as well as to clearly define its blueprint and technical documentation. Blueprints represent the assembly of a product. It involves all the key dimensions that are important for defining the location, integration, as well the product's own main dimensions. The given dimensions are important for manufacturing and installing such an assembly, while accessibility to these data should be restricted. A 3D model is of key importance as a presentation tool for direct customers, especially when the product is proportional to their expectations. A dispositional drawing is supposed to convince the customer by presenting the product's dimensions and thus presenting its size in space.

A product provides the performance of a required function in space (Fig. 13.3). A function is defined by the conditions of the transformation of energy, material and information. The existing understanding has not included spatial conditions, i.e., the product was dealt with in a non-volumetric way. With the advent of modelling, relations between the function and the shape have proved to be in strong correlation. The notion that a process can be performed without taking account of space, does not consider the real extent of the whole transformation process, performed by a particular product. Transforming information via electromagnetic radiation already requires a transmitter and a receiver, both in a specific size. For this reason, the blueprints conveying information about space and the size of a product should present all the important information about the position of the blueprint in the information chain of the entire R&D process, i.e., from the idea to the product.

13.2 Modelling Assemblies

Assemblies are modelled according to two procedures, mentioned in the previous sub-chapter. Assemblies require a specific approach, recognizable in modelers using a special approach: (1) building an assembly with an advance announcement (*Multibody part*) (2) an executed assembly, disassembled according to functions—forming new bodies that are included at a later stage.

13.2.1 Assembly Structure

Similar to real products, building blocks in modelling are finished components or sub-assemblies. So far, product presentations have been stored in separate files. For a welded assembly, body models and volumes are stored in a single file (*Multibody part*).

With assembly models, the objective is to integrate the geometric information (size of elements) into an assembly structure (components, sub-assemblies), and at the same time define the location of individual elements' files. This means that the elements' and sub-assemblies' files are together in a single folder, i.e., the assembly's folder (Fig. 13.4).

13.2.2 Bottom-Up Building of an Assembly

The bottom-up building of assemblies follows the principle that the created parts (products) are imported into one file, representing the whole assembly. It is important to define the internal relations. The components (products) are of course not modelled in the assembly file, as the elements of the assembly should first be dealt with under functional and shape conditions. Such completed models of the products, distributed

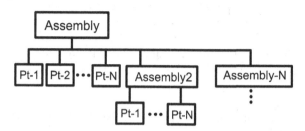

Fig. 13.4 A simple assembly scheme and the relations between the components. On the level 1 is Pt-1 and assembly 2 till assemly N. On the level 2 we have Pt-1 till Pt-N etc. Each level is defined according to the structure of the product

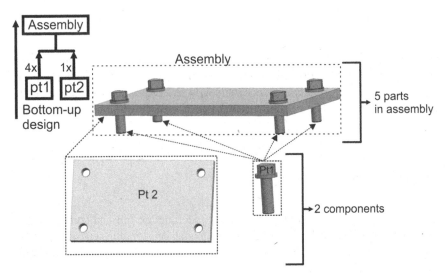

Fig. 13.5 Building blocks and their copies using the example of a panel and identical bolts

around the computer (usually in memory units), are then retrieved into a single assembly model and put together. It should be noted that the locations of the individual components in the memory must not change, but remain the same and be undeletable. The bottom-up process of building an assembly is graphically presented in Fig. 13.5.

13.2.3 Top-Down Building of an Assembly

For top-down building, an existing assembly and then supplemented with one or more components is used. The new components can be roughly defined by shape and can be given a function. They can also be standard, with a specific function, and these can be given blending functions, according to the required shape. The associativeness among the components can be in pure geometry, represented by the blending function. Alternatively, associativeness can be defined by a function, which is then again fulfilled by geometry, defined by the geometric conditions of the existing assembly. Figure 13.6 shows a simple assembly, consisting of two parts (*pt1* and *pt2*), where the second part (*pt2*) is built according to the top-bottom method. It was modelled after importing the first component (*pt1*). This was done by copying the geometry from the far edges of the *pt1* ears into the *pt2*. The parametric distance on *pt1* was defined, which can be randomly changed. A change of distance between the *pt1* ears directly affects the *pt2* geometry in the assembly. This is referred to as the second component, being associative to the first component. In principle, this is a defined relation and can then be optionally switched off.

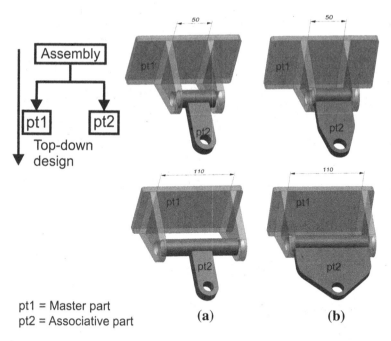

Fig. 13.6 Associations. (**a**) Dependence of one parameter, neglecting the influence of loads at an increased distance between the supports; (**b**) also shows the association for the second element, the fly jib, acquiring a new shape and an appropriate reshaping

The modelling principle, where the main component is called the **Master part**, and the others, the **Associative parts**, is called **Master Modelling**.

13.2.4 Relations Between the Components in an Assembly

In this case, an association is defined for the components that are imported into an assembly. The associations are exclusively geometric. At the beginning, each imported component should have six free degrees of freedom (three translations and three rotations). The relations among them can be defined between the existing objects as relative relations, or as absolute, relative to the global coordinate system. It should be noted that the global coordinate system is basically the coordinate system of the assembly, as this is the only way to upgrade the assembly across a wider area with a coordinate system that, in our case, becomes the local coordinate system. When defining relations, make sure to be at the level of the main assembly. In no way can you accurately define the relations in the local coordinate system for any of the components or in the space of a component that is accidentally identified with the global coordinate system. The sets of relations in assemblies (Fig. 13.7) in modern

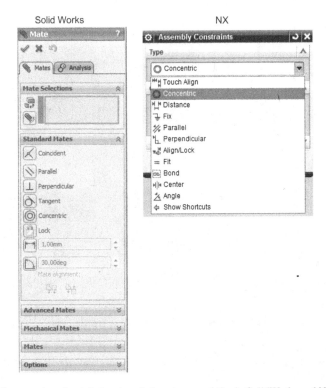

Fig. 13.7 Command set for defining the relations in assemblies in SolidWorks and NX software

modelers are designed in a very similar way, which means that they are identical in terms of functions. They can only have different names.

Figure 13.8 shows some basic relations, frequently found in modelers. Relations allow the linking of components by merging the geometric entities, such as points, lines (edges), surfaces, axially symmetric parts, etc. Some of them can be combined with one another. But in any case, the laws of geometry should be considered in order to prevent non-complementary relations between them.

13.3 Modelling in SolidWorks

13.3.1 Preparing Components

To model assemblies according to the bottom-up technique, you first need to prepare the components, independent of the assembly. The components are therefore ready for the bottom-up assemblies, only once they have been processed in all details. In this case, you should strive to process all of the newly developed components in all

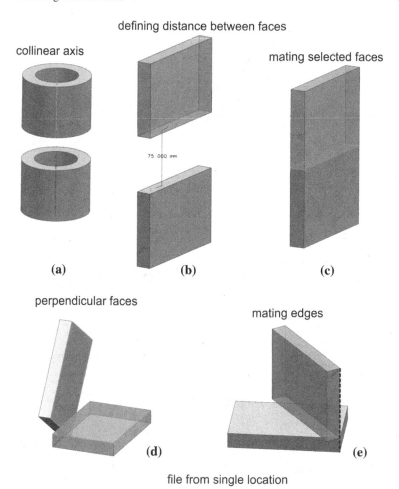

defining distance between faces

collinear axis

mating selected faces

75 000 mm

(a) **(b)** **(c)**

perpendicular faces

mating edges

(d) **(e)**

file from single location

Fig. 13.8 Some of the frequently used relations, that are found in modelers: colinear, concentric (**a**), distance (**b**), coincident (**c**), perpendicular (**d**) and mating (**e**)

details. Standard products or products accessible on the market in mass production can therefore be taken as quality defined elements, used to build an assembly.

13.3.1.1 3D Shaft Models

A shaft model (Fig. 13.9) can be acquired from previous records or can be created from scratch. The necessary data to create a model are shown in Fig. 14.50.

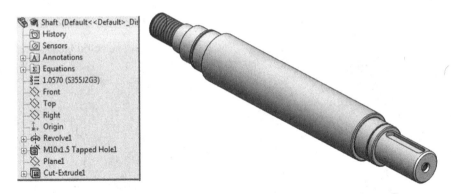

Fig. 13.9 Shaft model $\phi 60 \times 450$ mm

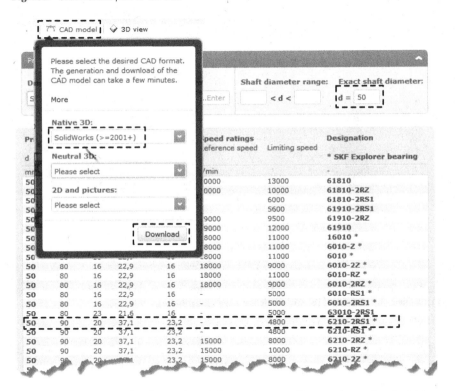

Fig. 13.10 Selecting a bearing and model transfer according to the enclosed database (www.skf.com)

13.3.1.2 3D Bearing Model

A CAD bearing model can be downloaded directly from the website of a producer. In our case, it is www.skf.com. In the bearing database, select one according to the load

Fig. 13.11 Selecting a model of the SKF 6210-2RS1 bearing with the dimensions $\phi90/50 \times 20$ mm

ratings (code 6210-2RS1). When exporting data for a CAD bearing model, make sure to select a suitable format (SolidWorks) and download the model (Fig. 13.10).

The model transfer is executed in the form of a *.swb* file. It includes the commands used by the software to generate the model. In our case, run macro (***Tools > Macro > Run***), select the downloaded file and generate the model. In our case, it is a bearing assembly, consisting of the inner and the outer ring, the left and the right seal and the rolling elements, i.e., balls (Fig. 13.11).

13.3.1.3 3D Model of Retaining Ring

A CAD model of a retaining ring can be taken from the software's internal library (***Design Library > Toolbox > DIN > Retaining Rings***). Figure 13.12 shows a model of the external retaining ring and Fig. 13.13 the internal one.

13.3.2 Assembly: Bottom-Up Design Technique

Having saved all the elements of the assembly, proceed to modelling the assembly. Open a new document (***File > New > Assembly***). A new assembly window opens. Continue by selecting the element that you wish to include first into the assembly. In our case, take the shaft as the basic element (Fig. 13.14).

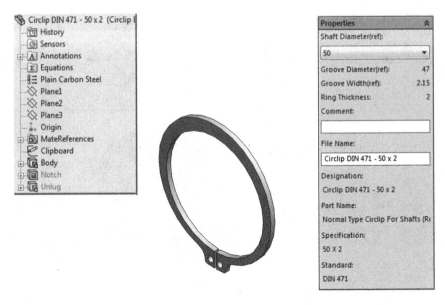

Fig. 13.12 Selecting a model of the external retaining ring $\phi 50 \times 2$ mm (DIN 471)

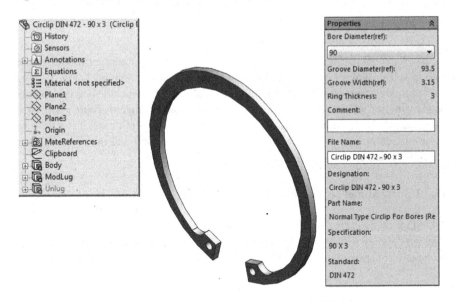

Fig. 13.13 A model of the internal retaining ring $\phi 90 \times 3$ mm (DIN 472)

SolidWorks offers three ways of positioning the parts in the assembly:

- **Fixed** (*Fix*)
  A part has an absolute fixed position in the assembly's coordinate system and cannot be moved.

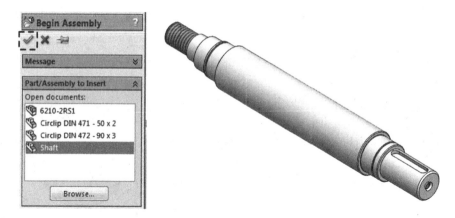

Fig. 13.14 Creating a new assembly and inserting the basic element—the shaft

- **Free** (*Free*)

  Not attached to the assembly, a part can be freely moved around the assembly
- **Relations** (*Mates*)

  A part is positioned by means of relations relative to the other elements of the assembly.

The first part to be included in the assembly is usually included with a fixed position because it is important for the modeller to be always positioned in its natural position in nature. As such, it provides the starting position, with the coordinates of the part's axes and those of the whole assembly coinciding. Each following element is usually positioned in the coordinate system by means of relations.

13.3.2.1 Inserting Components of the Assembly

Insert components into the assembly with the (***Insert > Component > Existing Part/Assembly, ...***) command or simply drag them into the assembly window (***Drag & Drop***) (Fig. 13.15).

Other components of the assembly are positioned by means of relations (***Mates***). The positions of the individual components are defined relative one to another, using the (***Insert > Mate, ...***) command.

- Concentric position of the bearing on the shaft (***Concentric***) (Fig. 13.16)
- Lateral mounting of the bearing onto the shaft (***Coincident***) (Fig. 13.17)
- Rotation restriction (***Coincident***)

  In principle, it is not necessary to restrict the bearing rotation; however, due to later manipulation it makes sense to fully define all the components. In this case, align the shaft's and bearing's front planes (Fig. 13.18).

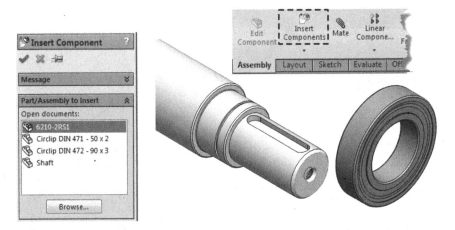

Fig. 13.15 Inserting a bearing into the assembly

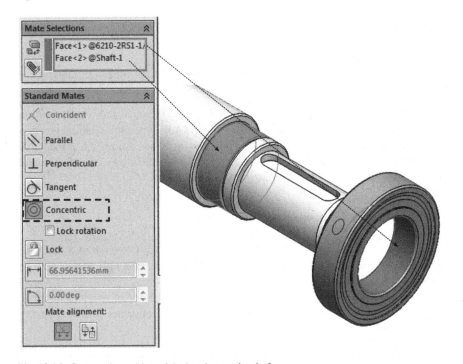

Fig. 13.16 Concentric position of the bearing on the shaft

Insert the retaining ring in a similar way as the bearing, paying attention to the axial position of the retaining ring in the groove. The retaining ring should fit on the outside of the groove, preventing the bearing from moving along the axle (Fig. 13.19).

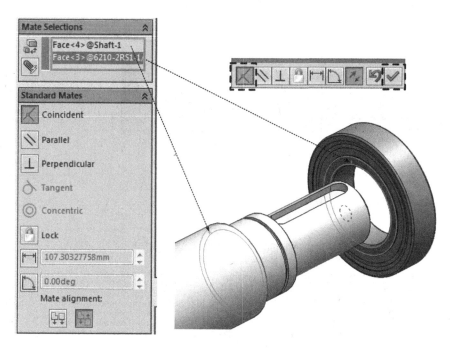

Fig. 13.17 Coincident position of the bearing on the shaft

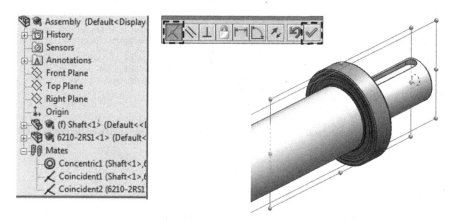

Fig. 13.18 Rotation restriction and assembly structure

When a component in the assembly is repeated, there are several ways of inserting several identical components in the assembly. The following methods are available:

- inserting a selected component again (***Insert, etc.***),
- copying a selected component (***Copy, etc.***),
- mirroring a selected component (***Mirror, etc.***),

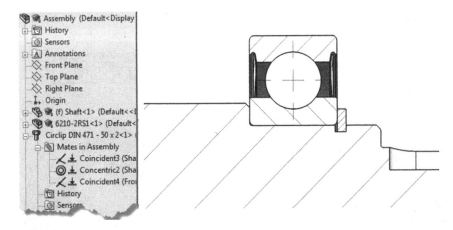

Fig. 13.19 Position of the retaining ring in the assembly

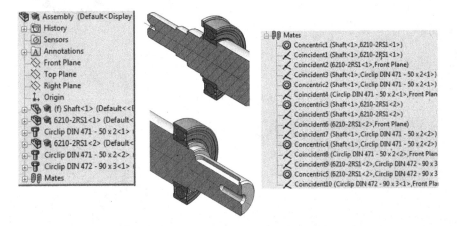

Fig. 13.20 A shaft, bearing and snap assembly using the bottom-up technique with the corresponding relations

- patterning a selected component (**Pattern, etc.**)

In our case, simply copy the selected components (the bearing and the snap) by selecting them and holding the CTRL key and then moving it into a new location. Having inserted a component, position it and save it in the assembly.

Now insert the internal retaining ring and position it (**Mate**) according to the bearing on the cutter side. Figure 13.20 shows a partial shaft bearing assembly using the bottom-up technique.

13.3.3 Modelling a Housing in the Top-Down Technique

Creating a new element in an assembly is similar to modelling an object, with the difference being the option to link new topological elements to the existing ones by means of relations. In our case, we will present the top-down technique to create a housing for a shaft bearing. The housing will be made out of a tube. For this reason, the dimensions of the outside ring of the rolling bearing will be taken as the basis. With its diameter ϕ90 mm, you need to plan for a tube with an outer diameter of at least ϕ100 mm. The seam tube database shows that the closest outer diameter of a standard tube is ϕ101.6 mm. The inner diameter should be long enough to provide an adequate surface quality. Plan for the tube wall thickness to be 8 mm. This can be achieved by high-quality machining of the diameter for the outside fitting of a ϕ90 mm bearing.

Write the modelling procedure for the housing on a piece of paper in order to be able to easily follow any discrepancies. You can also print out the axonometric model and write onto it the main housing dimensions, including the steps yet to be carried out. (1) Outside tube diameter of ϕ101.6 mm and wall thickness of 8.8 mm; (2) cut off to proper length; (3) create a sitting onto which; (4) legs are welded; (5) making bearing positions; (6) keyway grooves; (7) finishing (edge blending, chamfering, welds, etc.).

13.3.3.1 Positioning a New Element: A Housing

Inserting a new element into the existing assembly (***Insert > New Part***), it is first necessary to define a plane for the base sketch. This plane represents the front plane of a new part. Any of the existing planes can be selected. In our case, select the shaft's front plane from the tree structure (Fig. 13.21).

13.3.3.2 Tube Housing Revolve

The housing is created by defining the tube with the snap in the base sketch, rotated with the revolve feature. Tube parameters include: the outer diameter 101.6 and wall thickness 8.8 mm, with an extra 5 mm on the left and on the right (Fig. 13.22).

The tube-housing model is shown as a virtual element in the existing shaft bearing assembly (***Part Assembly***). Check whether the model has been properly fitted into the assembly and then you can save it as an external file with the (***Save Part (in External File)***) command. In order for clarity of the naming of individual models to provide their recognition, change the name (*Housing.sldprt*) and save it in the same folder as the assembly (**Same As Assembly**) (Fig. 13.23).

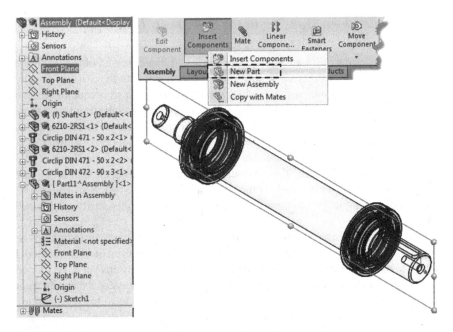

Fig. 13.21 Positioning the housing's front plane in the existing assembly

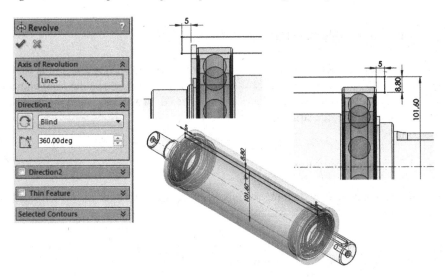

Fig. 13.22 Tube housing revolve $\phi101.6 \times 8.8$ mm

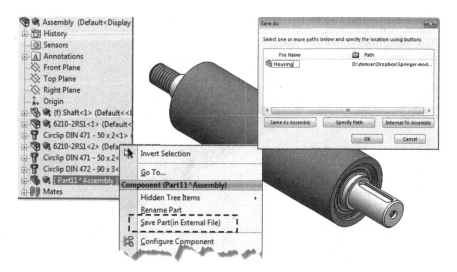

Fig. 13.23 Saving housing components within the assembly

13.3.3.3 Modelling Attachment Legs

The housing's attachment legs will be made out of flat steel with dimensions of 40×8 mm and a length of 140 mm. Define two bores of 12 mm for the M10 attachment bolts on both sides of the steel.

The structure for welding the attachment legs onto the steel tube is such that it requires aligning of the outside tube diameter first. It provides high-quality positioning and a good shape to form the welds. The flat steel will come onto the aligned cylindrical shape and it will be welded from the side with a fillet weld. Align the outside cylindrical part of the housing in the housing with the (*Edit Part*) command or separately, by presenting the element in its own box with the (*Open Part*) command. Remove the material by means of extrusion (*Extruded Cut*) (Fig. 13.24).

Proceed by adding legs (*Extruded Boss/Base*) with the merging bodies option (*Merge result*) switched off (Fig. 13.25).

13.3.3.4 Creating Bearing Positions in the Housing Tube and the Internal Retaining Ring Groove

The procedure is executed within the assembly using the edit component (*Edit Component*) command. During the execution, select a proper sketching plane first, and draw the bearing positions and the retaining ring. Having defined the shape, extend the sketch into space by means of revolving (*Cut-revolve*), using the existing geometry. With intersection edges (*Tools > Sketch Tools > Intersection Curve, etc.*) acquire

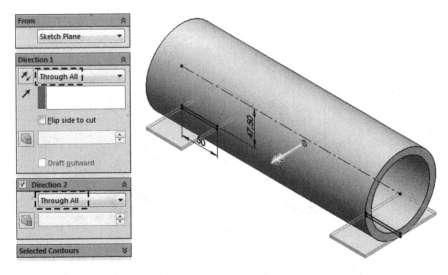

Fig. 13.24 Aligning part of the *circle* to prepare the welding plane for flat steel

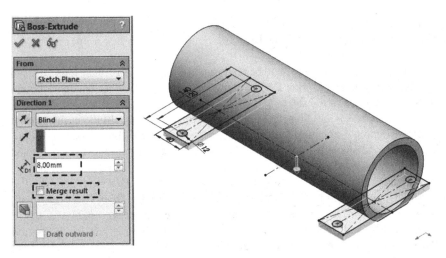

Fig. 13.25 Modelling attachment legs

the shape of the existing elements. It results in a shape and size adjustment according to the existing components (shaft, bearing, snap).

Using the (*For construction*) command, change the resulting contours into structural (support, in our case) ones. While modelling, pay attention to the functionality of the product at all times, and make sure that it can be manufactured and assembled. Continue by revolving the sketch (Fig. 13.26).

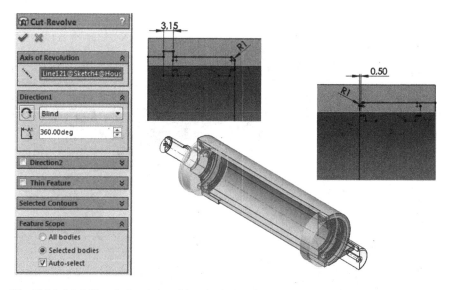

Fig. 13.26   Modelling the bearing positions in the housing

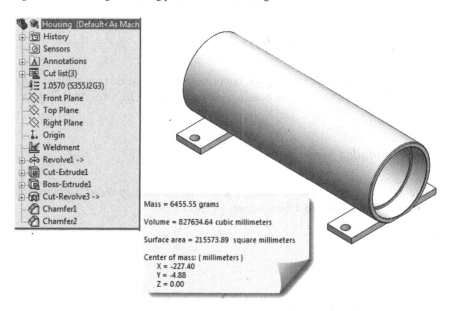

Fig. 13.27   Housing model with the corresponding features structure and mass characteristics

13.3.3.5   Adding the Chamfer and Final Shape

The housing is finished when the chamfers and welding points are completed (Fig. 13.27).

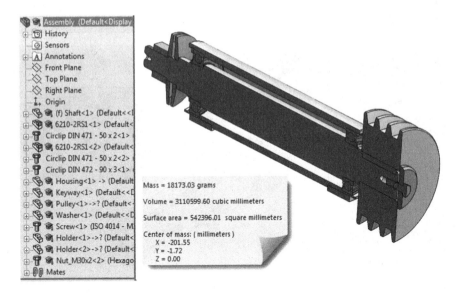

Fig. 13.28 Cross-section of the assembly model of a circular saw shaft bearing

13.3.4 Final Assembly Model

By adding other components (pulleys, cutter holder, keyway, etc) the shaft bearing assembly of a circular saw with a housing, shown in Fig. 13.28, takes its final shape.

13.3.5 Exploded View

Create an exploded view by creating a new configuration in the (*Configuration Manager > New Exploded View*) assembly (Fig. 13.29).

13.4 Modelling in NX

13.4.1 Bottom-Up Method

The bottom-up modelling of an assembly begins by opening a new file with the (*New > Model > Model/Assembly*) command. Siemens NX modelers use the same **.prt* file format for both parts and assembly models. The command sets for the assembly can also be freely activated for the model (Fig. 13.30).

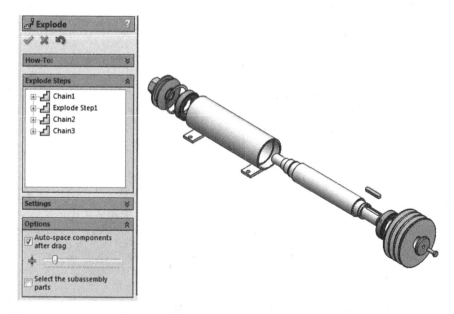

Fig. 13.29 An exploded view of the assembly

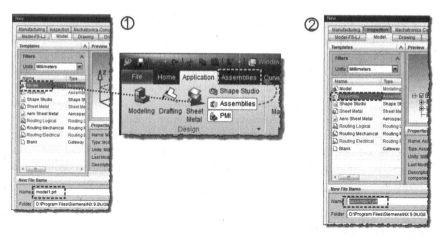

Fig. 13.30 In the bottom-up approach, open a file as a part modelling format and then switch to assemblies. In the case of top-down, the modelling begins in the assembly

All the commands are accessible in both approaches and you should opt for the steps, typical of step-by-step, bottom-up modelling.

The assemblies are modelled by including the individual parts, elements or modules into the main, i.e., the assembly, file. Locate each part of the model into the assembly file space and then define the relations among them. In this approach it is typical to first specify the details of a particular part and then merge the parts

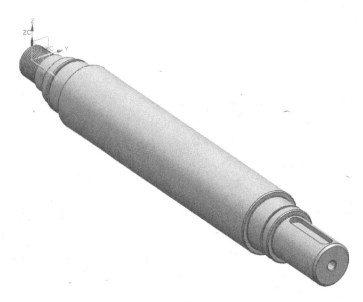

Fig. 13.31 A pre-modelled CAD shaft model ϕ60 × 450 mm

in the assembly file. It is logical that some predictions of the shape that are suitable and defined on the element's module in advance, should be modified in the assembly, where they are merged and also optimized in terms of the shape. Better recognition of the required shape details while modelling parts of the assembly means fewer modifications and shape adjustments in the assembly. This method tests the designer's ability to understand both the technologicality and the forming shape details. Due to the volume of information it is recommended to make a paper copy of a rough shape, followed by sketching detailed shapes onto the paper. Only then are harmonized shapes transferred to a modeller in order to model a product. This also allows traceability of a changing shape later, during subsequent procedures and verifications.

13.4.1.1 3D Shaft Model

A shaft model (Fig. 13.31) is modelled according to the procedure, presented in Chap. 6, dealing with revolve. Having finished modelling, save the file under a new name in an accessible place. A new folder for each new assembly is recommended.

13.4.1.2 3D Sub-assembly of a Standard Bearing

The producers of standard machine elements are interested in providing designers with pre-constructed 3D models. This allows the designer to quickly insert a CAD

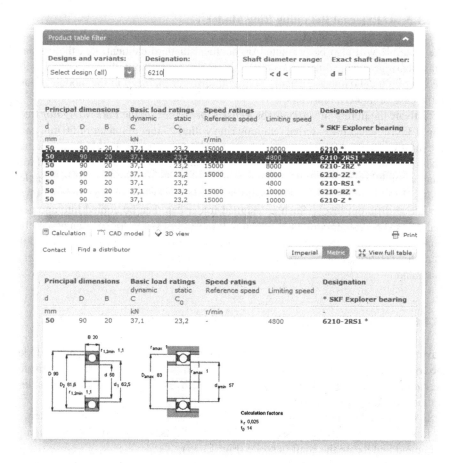

Fig. 13.32 Standard bearing generator, example: 6210-2RS1 (www.skf.com)

model into his or her design assemblies, i.e., complex products. The designer registers with a particular webpage and chooses a product, whose geometry is also presented as a 3D model. Download the selected model to your working environment, i.e., a database on your computer.

An example will be carried out with the assistance of an on-line bearing generator at www.skf.com. A similar method could also be applied to the data transfer from a disc, a local database, etc. Find a part—a bearing in our case—in the table according to the pre-set parameters (Fig. 13.32).

Many generators allow the transferring of files for generic models in formats directly for a particular modeller. In our case, select the NX7 format or later. Usually, there are also general neutral formats available, such as *.step* or *.iges*.

13.4.1.3 3D External Retaining Ring Model

The example will also be presented on another standard element, i.e., the extrnal retaining ring with the dimensions $\phi 50 \times 2$ mm. In this case it will be presented with the use of a local database, usually built with discs or included in the modeller—as in the case of modern modelers.

Activate the library in vertical tabs *Reuse Library* next to the *Part Navigator* tree (Fig. 13.33). Select the external retaining ring with the inner diameter of 50 mm and thickness of 2 mm. After checking the presented data, save it in a folder in a specific location. When the file locations are not specifically defined, it is recommended for small assemblies to save all their component files and the assembly itself in a single folder. With complex assemblies, build the model according to the principle of a module at different levels. Such a designed folder can then be copied into another folder or even another computer, which will make it possible for such an assembly to establish new upgraded relations. Having transferred the data to a new environment, the relations can be restored using the (*Assemblies > Replace Component*) command.

13.4.1.4 Creating an Assembly: Transferring 3D Models to the Assembly

Having generated the components, go back to the assembly file. It does not include the commands for forming geometric entities or features. In the main menu, activate the *Add Component* tab (Fig. 13.34) and continue by adding larger and then smaller components into the assembly.

In our case, the shaft is dimensionally the largest component, so it goes in first.

Include the shaft into the assembly file with the (*Assembly > Add Component*) command and locate the shaft in space (Fig. 13.35).

Having placed the shaft in the assembly file, proceed by defining the degrees of freedom for the object (shaft), placed in space.

Initially, the shaft, placed in the assembly's space, has all six degrees of freedom (DOF) undefined. The first task is to define all of the six degrees of freedom in order to define the fixed origins, first for the shaft and indirectly for all the other components that are subsequently placed in the assembly's space. The procedure is executed with the (*Assembly Constraints > Fixed*) command. In our case, do it by determining a value for each degree of freedom, which is then inserted with the (*degrees of freedom—DOF*) command (Fig. 13.36).

13.4.1.5 Transferring Bearings to the Assembly

There are also bearings in the shaft assembly. They are also transferred to the assembly according to their orientation and position. With the shaft supported by two bearings in our case, one will be (original 3D model) transferred to the assembly and the other will then be copied within the assembly. A bearing is a part, composed of several

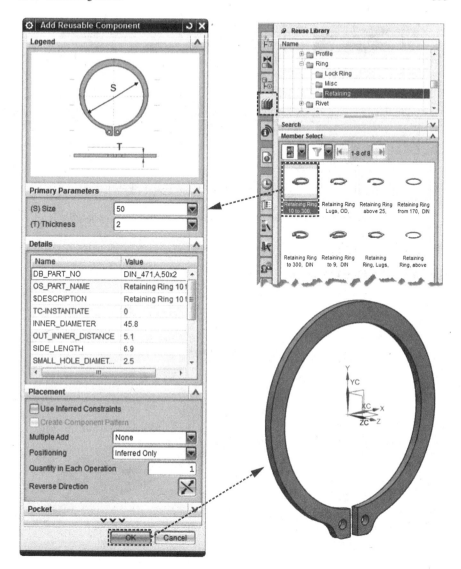

Fig. 13.33 Importing the external retaining ring $\phi 50 \times 2$ mm into the assembly

components and thus a sub-assembly in its own right. Due to this fact, the bearing assembly file must be selected, not its components' one. The components should be attached in the same folder because the data are transferred together with the components, specified in the same folder.

When transferring a 3D bearing model into the assembly, there are no specific locations or orientations and all six degrees of freedom are free. Place the 3D model

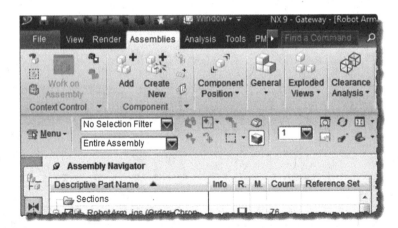

Fig. 13.34 Interface environment for working with assemblies

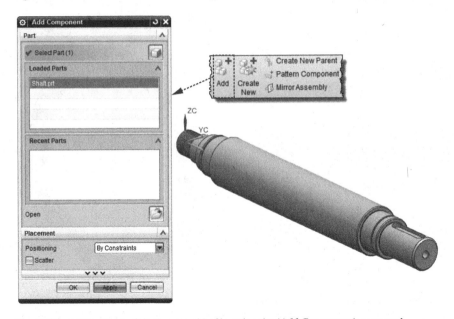

Fig. 13.35 Adding the shaft in the assembly file, using the (*Add Component*) command

near the shaft, i.e., near the first component, transferred to the assembly's space (Fig. 13.37).

Determine the bearing's position by specifying the location and direction constraints for each degree of freedom. As the location (x, y, z) and orientation (α, β, γ) within the assembly had already been set, the freely located and oriented bearing must be attached to the shaft by relations. The procedure is executed with the (*Assembly Constraints*) command. In our case, first select the concentric positioning

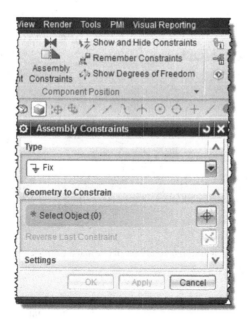

Fig. 13.36 Defining the position of the shaft in space

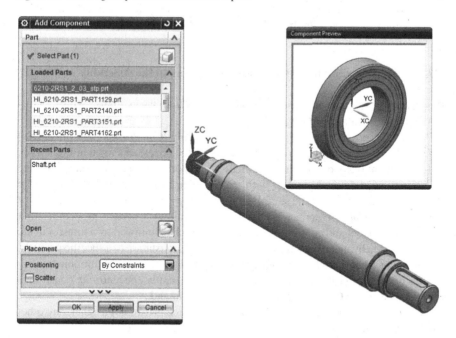

Fig. 13.37 Inserting a bearing into the assembly

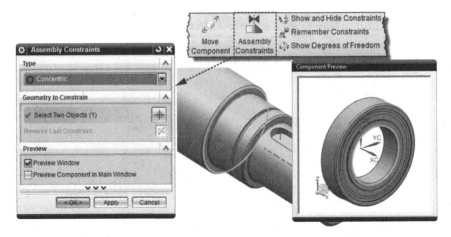

Fig. 13.38 Relations between the *circles* and axes of one and the other element. Relations between the shaft and the bearing

(*Concentric*) and select the circles on the edges. They are characteristic of both parts and define the links between the locations, as shown in Fig. 13.38.

The presented relation allows confirming the relation with the circles in the plane and the centric relation with their axes.

13.4.1.6 Transferring the External Retaining Ring to the Assembly

The external retaining ring is transferred—imported into the assembly using a familiar (*Concentric*) command (Fig. 13.39).

Apply the position setting with the (*Assembly Constraints > Concentric*) command and use a relation to link the circles on the edges, as shown in Fig. 13.40.

It was mentioned before that the shaft is supported by two bearings. As one has already been defined together with the retaining ring, the other bearing will be placed on the other side of the shaft by means of copying. To do that, use the (*Move Component > Copy*) command. The execution follows a certain procedure: first define a component and confirm it (Fig. 13.41), which is followed by a transfer in the assembly space by means of automated copying and translation to a specific position. Repeat the procedure for the external retaining ring and all the other components. For reasons of accuracy for the positioning and centring, apply concentricity constraints with the (*Assembly Constraints > Concentric*) command.

Because all the components are not bound by concentric relations, there are other possible relations, such as (*Assembly Constraints > Touch Align*) (Fig. 13.42). Transitions and translations—together with rotations that form the relations between the characteristic geometric parameters—can be combined on the same part or element.

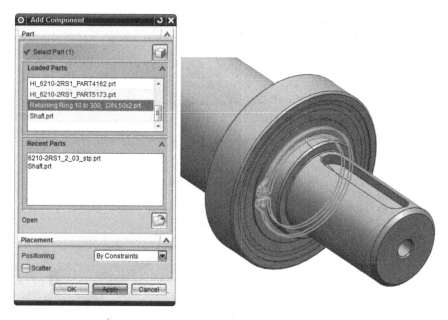

Fig. 13.39 Retaining ring transfer to the assembly's space

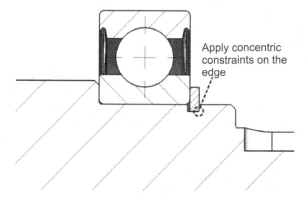

Fig. 13.40 An important detail of positioning the retaining ring function in the groove relative to the bearing. Groove backlash is set by a standard

When establishing relations, make sure to not contradict the rules between the relations.

After transferring all the parts and elements to the assembly space, it is important to check before the end of the bottom-up modelling, the situation with respect to specifying the degrees of freedom, i.e., where there is no specified degree of freedom for a particular element or part. Controlling or checking, to be more precise, is performed by right clicking (**MB2**) the assembly tree (**Assembly Navigator**) (Fig. 13.43). A new menu opens and (**Show Degrees of Freedom**) should be selected (Fig. 13.44). A 3D

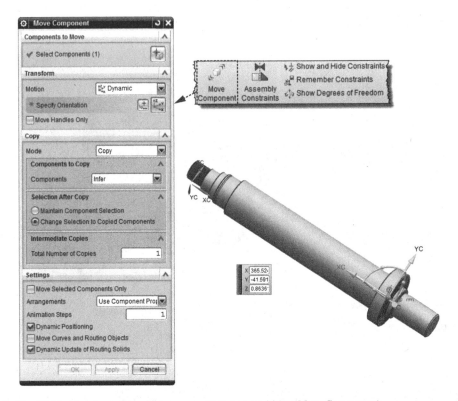

Fig. 13.41   Copying and moving components to new positions (*Move Component*)

model will then display the indication of undefined degrees of freedom, which can then be hidden by a right click and the (*Refresh*) command.

13.4.2 Top-Down Method

This method allows establishing the associativeness between the elements in an assembly. It is useful to supplement assemblies with new elements, or elements, determined by parameterization. The advantage is that when supplementing one component or when changing its shape, its associated component will change, too.

Below, we will use the presented assembly of a shaft with bearings and retaining ring in such a way as to supplement it with the housing and the internal retaining ring. It should be noted in this case that adding the retaining ring it first requires the creating a sketch with a functional link between the bearings of the internal retaining ring and the housing. It is only the sketch that makes you understand the logics

Fig. 13.42 A list of possible relations between components in the assembly

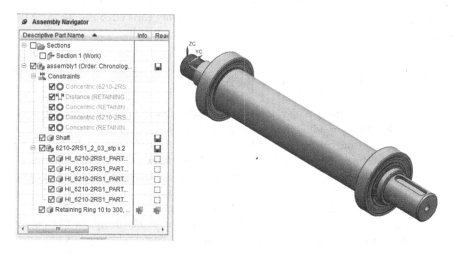

Fig. 13.43 A finished assembly, following the bottom-up method

and link between the presented elements. The internal retaining ring therefore goes into the assembly in advance, as it significantly reduces the processing time for the housing, where the internal retaining ring groove should be taken care of.

Fig. 13.44 Degrees of freedom on a part or an element in the modelled assembly

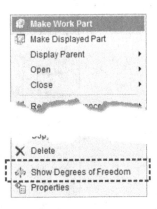

13.4.2.1 Adding the Internal Retaining Ring ϕ90 × 3 mm

Use the library of standard elements and find an internal retaining ring with an outer diameter of 90 and a thickness of 3 mm. Transfer the part to the assembly space and define its location with the concentricity relation (*Concentric Constraints*). As a relation, use the short arc at the internal retaining ring ear's edge and the edge of the bearing (Fig. 13.45).

13.4.2.2 Modelling the Housing in the Assembly

The assembly is complete with components that are sufficient in a particular part. But only the housing for guiding the bearings will complete all the necessary elements or parts. The housing can be modelled as a part that is to be included in the assembly. The assembly as a new part is created with the (*Assemblies > Create New*) command (Fig. 13.46). The newly created part should be named, Housing, for example. For the initial data, it is important for you to be at the assembly level. Once the bases for defining a part at the assembly level have been set, you can proceed to the part level. This is done with a right-click (*MB2*) in the assembly navigator (*Assembly Navigator*). A new window opens. Select *Make Work Part*. Following this sequence of commands will bring you to the part level (Fig. 13.47). The inactive parts of the assembly become transparent on the screen.

On the XC-ZC plane, you can sketch the housing (Fig. 13.48). It will be created by means of revolving (*Revolve*) around the shaft's main axis (Fig. 13.49).

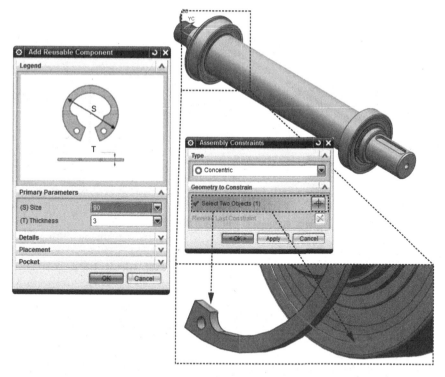

Fig. 13.45 Transfer and placement of the internal retaining ring $\phi 90 \times 3$ mm

Fig. 13.46 Defining a new
component in the assembly
space

13.4.2.3 Modelling Housing Supports for the Attachment Onto a Surface

Create cut-out for two seats on the bottom of the housing applying a simple extrusion
and the removal of material(Fig. 13.50) (**Extrude**).

The supports are very simplified because the loads are not high and the alignment on the outer diameter determines their positioning seat. Create the sheet metal

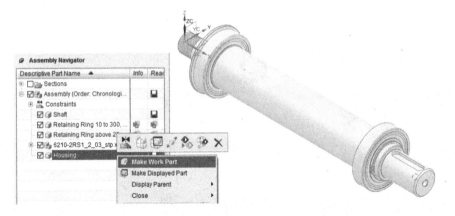

Fig. 13.47 Transfer to a working component within the assembly space

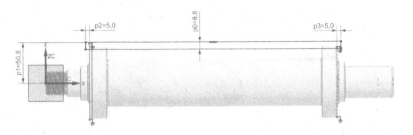

Fig. 13.48 Modelling the housing sketch

supports by means of extrude (**Extrude**). Create the assembly with the **Boolean** command, using the low dash to create the assembly. Using the **None** command (Fig. 13.51), create three separate bodies on the housing, similar to the modelling of welded structures. In our case, the supports are also welded on the cylindrical housing body. On both supports, create four ϕ12 mm attachment bores over a distance of 120 mm (Fig. 13.52). Product is simple and that is the reason that the supports are simplified as well. Supports are from profile who is created with command (**Extrude**).

13.4.2.4   Creating Detailed Shapes in a Tubular Housing to Fix the Retaining Ring and Bearings

The presented housing model still lacks all the necessary functional attributes for direct use. The presented parts and elements possess all the key elements of functionality, they only lack the direct installation of internal retaining ring into the grooves, which need to be inside the housing. In this section of the modelling, you are at the level of the part *Housing*. To present the entire environment where the retaining ring will be installed, i.e., the housing, the surrounding geometry of the assembly will be copied into the part. In this space, it will be used to define the linking shapes

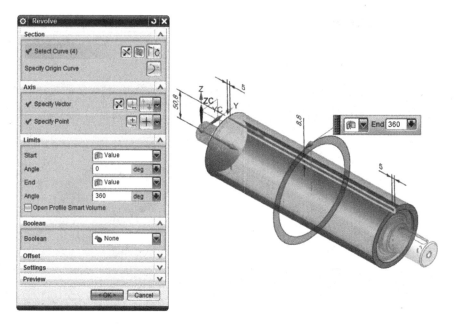

Fig. 13.49 Modelling the housing by means of revolving (*Revolve*) the base sketch around the shaft's axis

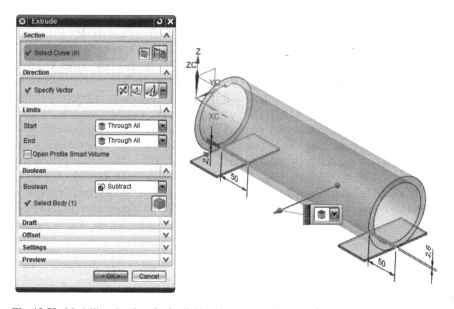

Fig. 13.50 Modelling the chamfer for linking the support sheet metals

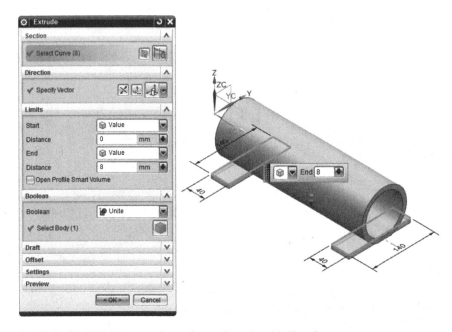

Fig. 13.51   Modelling elements from a flat profile on a welded housing

and functionalities, which can then be hidden, if necessary (***Show/Hide***). Using the ***Assemblies > General > WAVE Geometry Linker*** (Fig. 13.53) command, create the linking shape functionality (associativeness) with the geometry of the cover. You can choose from different geometric entities (curves, surfaces, bodies). In our case, copy the edge surfaces of the snap and the bearings, whose dimensions interfere with the body of the housing (Fig. 13.54).

Having presented the geometry of all the linked models of the components, you can approach the modelling a new sketch on the mid XC-ZC plane. Use the sketch for modelling a new rotated body, used for removing material in the housing (***Revolve > Subtract***) (Fig. 13.55).

Following the activities at the level of the part, go back to the assembly navigator (***Assembly Navigator***) to the higher level and activate all the components (Fig. 13.56). If necessary, include a new element or a part with extra modelling, following the described procedure (Fig. 13.57).

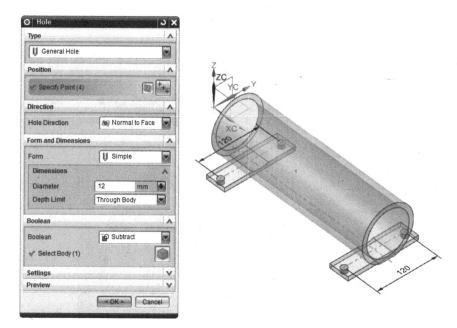

Fig. 13.52 Modelling four $\phi12$ mm bores over a distance of 120 mm

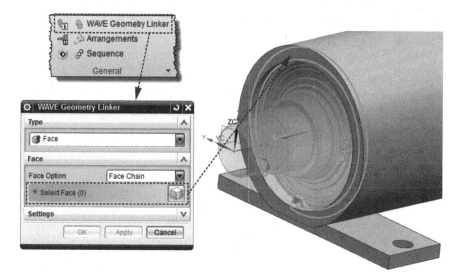

Fig. 13.53 Creating a relation from the snap's outside geometry in the assembly of the housing's inner shapes

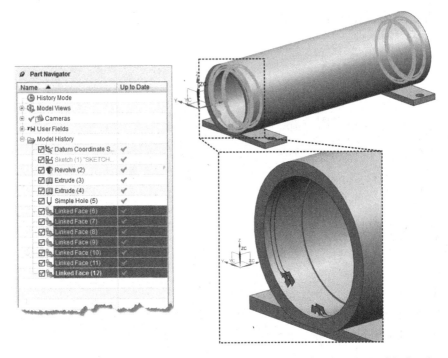

Fig. 13.54 Modelling the parts of bodies whose dimensions interfere with the body of the housing

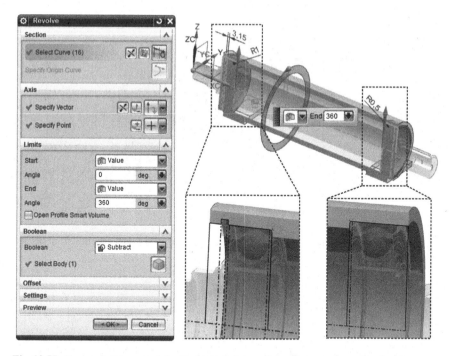

Fig. 13.55 Removal of the material with detailed shapes of two elements (snap, bearing) by means of rotating (***Revolve***) the base sketch about the shaft's main axis

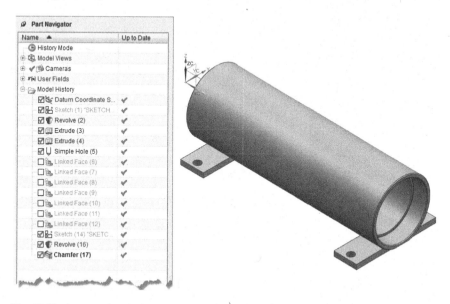

Fig. 13.56 A comprehensively modelled housing is executed at the part level

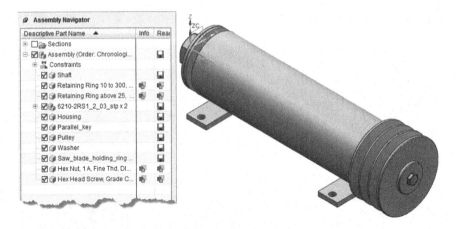

Fig. 13.57 A complete model at the assembly level with a few additional components, not specifically defined

13.5 Examples

Figures 13.58, 13.59 and 13.60 present few examples of asseblies.

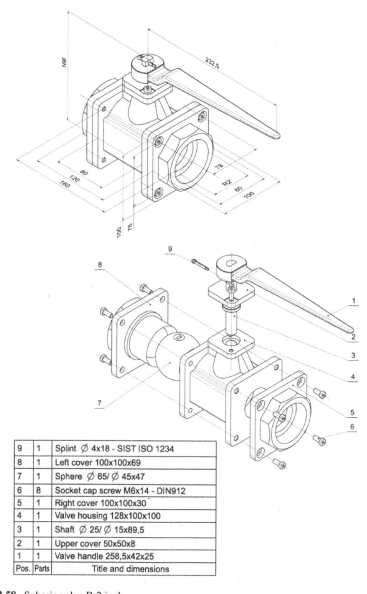

9	1	Splint ∅ 4x18 - SIST ISO 1234
8	1	Left cover 100x100x69
7	1	Sphere ∅ 65/ ∅ 45x47
6	8	Socket cap screw M6x14 - DIN912
5	1	Right cover 100x100x30
4	1	Valve housing 128x100x100
3	1	Shaft ∅ 25/ ∅ 15x89,5
2	1	Upper cover 50x50x8
1	1	Valve handle 258,5x42x25
Pos.	Parts	Title and dimensions

Fig. 13.58 Spheric valve R 2 inch

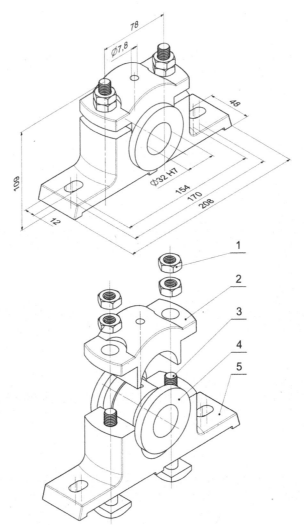

5	1	Lower housing 208x62x48
4	1	Bearing liner ⌀ 48/ ⌀ 32x64
3	2	Square head screw M12x98
2	1	Upper housing 108x48x47,5
1	4	Nut M10x1,5 - SIST ISO 4035
Pos.	Parts	Title and dimensions

Fig. 13.59 Bearing with housing $\phi 32 \times 64$ mm

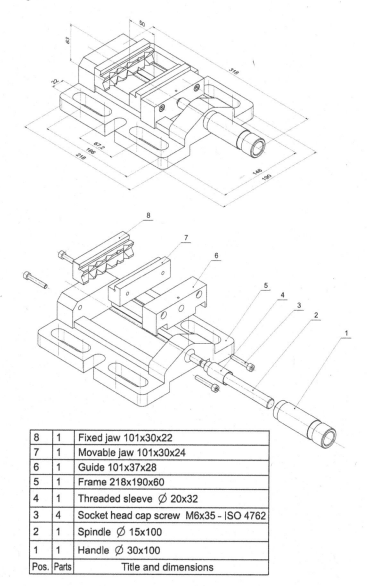

8	1	Fixed jaw 101x30x22
7	1	Movable jaw 101x30x24
6	1	Guide 101x37x28
5	1	Frame 218x190x60
4	1	Threaded sleeve ⌀ 20x32
3	4	Socket head cap screw M6x35 – ISO 4762
2	1	Spindle ⌀ 15x100
1	1	Handle ⌀ 30x100
Pos.	Parts	Title and dimensions

Fig. 13.60 Hand mechanical vice 100×50 mm

Chapter 14
Technical Documentation (Drawing)

Abstract This chapter presents detailed techniques for both modelers, necessary for a high-quality presentation in accordance with the standards of technical documentation. The goal was to present the problems that any designer can encounter, and then the tools to overcome them by means of different techniques in order to arrive at the end product, i.e., the technical documentation (manufacturing drawing). Also presented is a technique—probably not yet perfected—of using both modelers to create detailed working documents, which will require more accurate programming in the future. However, such information is very useful and necessary for the user in order to save him or her from searching for a suitable menu.

14.1 Assembling Drawing

Technical documentation is an important part of the engineering description of products that are manufactured using a variety of familiar technologies. It represents a graphical and descriptive representation of a product in all its detail. The object in space is shown as a graphical representation in the form of an assembly drawing or a common drawing, while the details are specified in a manufacturing drawing. More recently, graphical representations have been supplemented with photographs, especially in the case of the dynamic phases of the process that are managed by the product. Some parts of the manufacturing process require the assistance of a dynamic picture. Here, film, video or any other recording that captures and presents moving processes and procedures, and clearly presents various assembly and disassembly procedures, can be used.

Besides a graphical representation, descriptions of specific technologies, procedures and details are also important. Descriptions usually include calculations, material tests, creating specific technologies, etc. In this part, the focus will only be on the graphical part that is generally presented by assembling and manufacturing drawings.

J. Duhovnik et al., *Space Modeling with SolidWorks and NX*,   413
DOI: 10.1007/978-3-319-03862-9_14, © Springer International Publishing Switzerland 2015

Table 14.1 An information table, providing the description of an element, a first-level building block, shown in a manufacturing drawing

Manufacturing drawing which represent the all informations for users	
1. Details definition of the form	
2. Economy of details and their manufacturing technology	
3. Documentation value for description	
4. Description of pieces (numbers, differences and combinations)	
5. Design changes and capabilities	
6. Disabling repetition	
7. Rough definition of material (detail definition are on the technology documentation)	
8. Manufacturing principle:	(a) sequence operation
	(b) tools request
	(c) specific operation plan (gruops, flexsible, sequency)
9. System of the drawings or their complex	
10. Drawing technology (drawing technique, modeler used—file form)	
11. Product description and its document for the archive	

The vital information that a manufacturing drawing should include is presented in Table 14.1.

Using the existing technique of expressing and—most of all—specifying the representation layers, the manufacturing drawing is presented layer by layer. Each layer provides a separate approach for each user and thus ensures the carefully protected industrial knowledge of a particular environment. Specific technologies or typical processing must not be described in generally assessable documentation. By activating different layouts of the drawing, the level of accessibility for a particular user will be determined, as shown in Fig. 14.1.

Similar to the importance of introducing features into the development of modellers and the acceleration of modelling, the next step to define more clearly the transition from a model to technical documentation will be even more important.

A digital model requires data transfer to the forms, suitable for technical documentation, called a plan. For example, the simplest of elements, a bolt, is presented as a solid body in a cross-section in a broken-out part of a shaft, if technical drawing and standards are considered (Fig. 14.2a).

However, if the same requirement for a shaft cross-section is mapped to the technical documentation, it results in the representation as shown in Fig. 14.2b. Technical drawing rules allow a clearer expression of details and specific shapes that significantly contribute to a faster understanding of the whole complex. A similar difference is shown in Fig. 14.3. Its purpose is to show how careful you need to be with the direct use of incomplete projections and not respecting the standards.

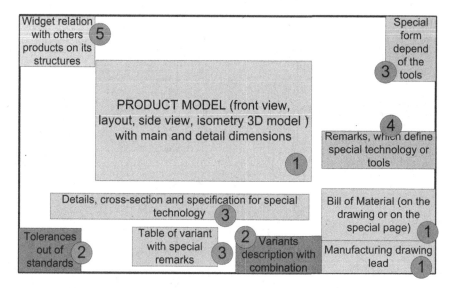

Fig. 14.1 Presenting drawing layouts and their levels for direct user contacts

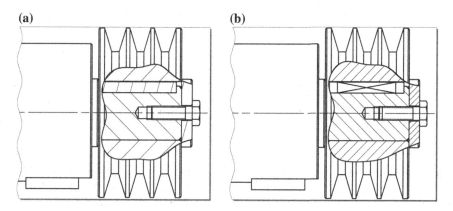

Fig. 14.2 Cross-section of a bolt, screwed into the shaft: representation directly from the modeller (**a**) and according to the ISO standard representation (**b**)

Transferring a digital model should be considered as a direct use of standards with the use of a high-quality expert system. Standardization conditions should be understood in the sense that the representation of a model should be clear in both cross-sections and views, and details and dimensions cannot be ambiguous (Fig. 14.4). The very same requirement also applies to the accelerated use of modellers. It includes not just modelling, but particularly the accelerated creation of manufacturing drawings, and—in the case of the bottom-up system—also a complete representation of a complex product assembly according to the rules of technical drawing. We as engineers expect that the creators of this second part will use more engineering skills for

(a)   SECTION C-C   **(b)**   E-E

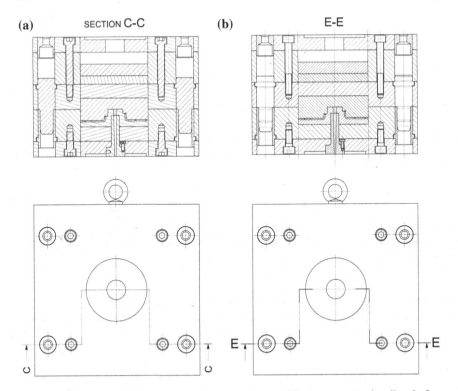

Fig. 14.3 The main part of the tool for plastic injection moulding **a** representation directly from the modeller, **b** conforming to the ISO standard

presenting the shapes with greater complexity, and speed up the creation of appropriate expert modules for various standards across the world. It is also expected that they will take account of those developed environments where technical knowledge has been developed and upgraded to top engineering performance, since knowledge is required there for the purpose of a clear representation and understanding.

Later in this chapter, a couple of typical examples of manufacturing documentation will be presented. Attention will be drawn to the additional steps, required for one or the other modeller. In any case, creating technical documentation without a detailed knowledge and familiarity with the philosophy of representing elements and assemblies cannot be used directly. Current development in this area is poor and it will take a lot of effort to upgrade and improve modellers. A designer not following the instructions of a particular modeller for representing elements and assemblies and not upgrading them with a high-quality understanding of technical drawing standards can expect major complications or at least a partial non-understanding in terms of presenting the documentation.

The argument that all participants should accept an unclear representation of plans is wrong in it its own right as it was the representation standards that provided a significant advantage in the international exchange of engineering projects.

Fig. 14.4 Tool for plastic injection moulding is well present in explorer view

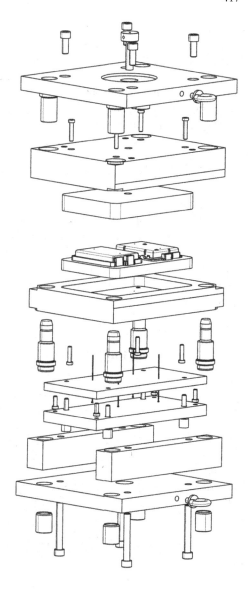

The chapter is designed as a preparation for the comprehensive process of creating technical documentation with the original 3D model shape. A special emphasis will be on the quality of the assembly and the manufacturing drawing. The elements that a particular type of documentation should include, have already been made familiar to students during the technical documentation course, and will only be adopted in this part. The exercise mostly includes a presentation of the different methods for creating individual elements of the documentation (view, cross-section, detail, etc.).

Modellers are not usually accurately set for the direct use of ISO standards for displaying and printing plans. With the engineering culture of sellers in the area of creating high-quality technical documentation being very low, we need to do the setting ourselves. In large companies, this is usually provided by a special service. It is wise to do the pre-setting only once. In our exercises, the problem will be presented by pointing out the errors.

Attention will be drawn to the most common mistakes. One of those is the hatching in cross-sections. We believe that the mistakes are a result of a misjudgement that serious software can be generated or—to put it better—produced anywhere in the world in order to make it as cheap as possible. This is why the technical culture from the environments with deficient or indefinite ways of expressing is making its way into modelling software. The status of each modeller should be checked to see whether it is suitable for the direct use of international standards. If it is not properly set, make sure to pre-set the modeller to the legally prescribed standards. Do this with the assistance of the retailer, who has an obligation for a specific market and the legislation of a particular country where the modeller is being sold (ISO, BS, DIN, SIST EN). It should be noted that a modeller provider should be warned to appropriately update the version for sale in the market of the European Union.

14.2 Manufacturing Drawing

Presenting manufacturing drawing (objective, information included, etc.) and special cases (welded piece drawing, sheet metal products, etc.) (Figs. 14.5, 14.6).

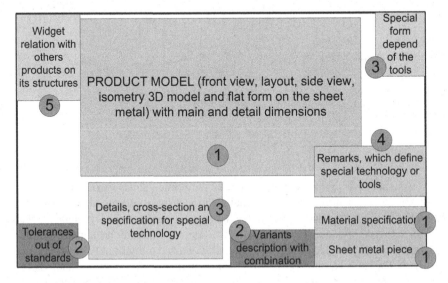

Fig. 14.5 Manufacturing drawing for a sheet-metal piece

(a)

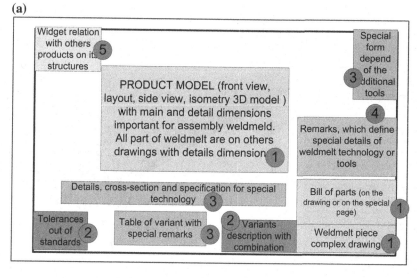

(b)

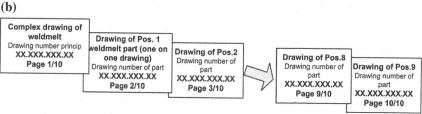

Fig. 14.6 Weldmelt piece: complex drawing (**a**), numbering principle by weldmelt piece with 9 parts for welding (**b**)

14.3 Modelling in SolidWorks

Begin by opening the 3D model that you wish to present in the form of technical documentation. In our case, create the assembly drawing first. To do this, open an existing assembly (*Assembly.SLDASM*) (Fig. 14.7).

14.3.1 Drawing Sheet Format

A new set of plans or models should begin with a new drawing (*File > New > Drawing*), which requires a new sheet format. In principle, all modellers include a set of default sheet formats; however, they usually do not conform to the ISO standard and should therefore be corrected or updated. You can create your own sheet formats according to your own specifications or the company's internal standards. In our case, the *A3 (ISO)* (Fig. 14.8) sheet format will be used.

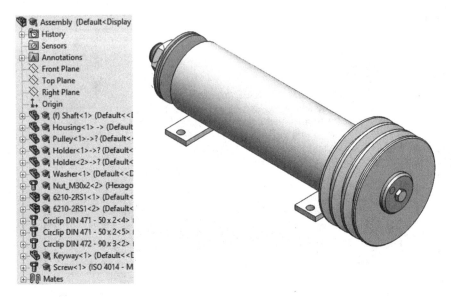

Fig. 14.7 Assembly model of a circular saw shaft bearing

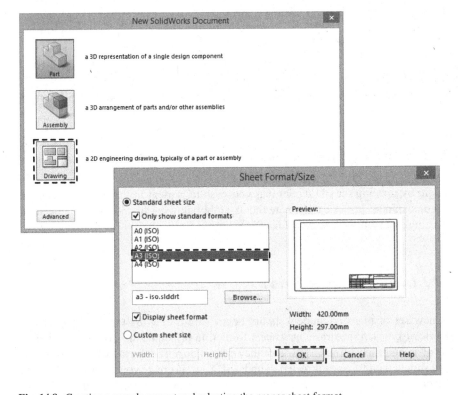

Fig. 14.8 Creating a new document and selecting the proper sheet format

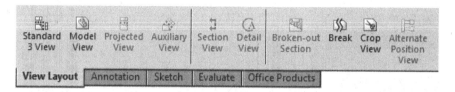

Fig. 14.9 The set of commands for creating views in the drawing

14.3.2 Views

When creating technical documentation it is important to select the proper views (*Insert > Drawing View, ...*) (Fig. 14.9). The number and type of views depend on the particular model and the information to be presented in the drawing. For this reason, it is often important to first sketch a rough concept of views and cross-sections, and then try to define them in the drawing space in terms of both size (scale) and layout. The size of a particular part of the drawing and the particular layout provide a conceptual form of the plan on a selected or frequently used standard format.

14.3.2.1 Model View

Orthogonal views that represent a model are defined by the use of the (*Insert > Drawing View > Model, ...*) commands. For each view that provides a high-quality representation of the model, it is necessary to clearly define its purpose and the required quality. The model should first be viewed in the 3D environment. Top, front and side views are determined by looking at its basic shape and its position in space. If a model has already been opened in the modeller, you can select it in the open documents window. Otherwise, find a model with the (*Browse*) command from the existing files. Having chosen a model, you can switch to the new window and select the view (top view in our case) and—using the mouse—position the view in the drawing (Fig. 14.10).

Centrelines and other adjustments to a plan in SolidWorks are inserted at the end, while reviewing the plan and checking its conformity with standards. This explains why the model lacks centrelines and other characters at the beginning, when you open a file that represents a model.

14.3.2.2 Projected View

There are several methods for creating the different views and projections of a model.

1. The easiest way is to use the *Standard 3 View* command to specify the front view for our model, and the projections (top and side views) are made by the software automatically.

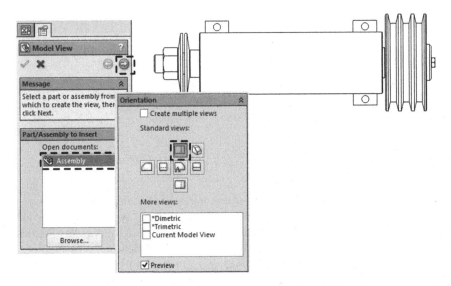

Fig. 14.10   Creating an orthogonal view. In this case it is a top view of the model in the drawing

2. Insert an orthogonal view (***Model View***), and move the mouse in various directions in order to insert the projections of the selected view.
3. The third option is to use the ***Projected View*** (***Insert > Drawing View > Projected***) command. In this option, select a suitable view in the drawing, followed by moving the mouse in the directions where different projections are displayed. Insert your projection of choice into the drawing by clicking on the viewing point (Fig. 14.11).

14.3.2.3 Broken-Out Section

A model and its details are best presented by means of cross-sections and broken-out sections. You can see that SolidWorks does not allow the creation of broken-out sections on cross-sections. Instead of the front view, you need to create a broken-out section (***Broken-out Section***) across the whole view. To do this, first create a closed contour (rectangle) across the whole view, followed by activating the ***Broken-out Section*** (***Insert > Drawing View > Broken-out Section***) command. When creating a cross-section on an assembly, a dialogue box opens, where you can specify the component that you do not wish to present in cross-section (e.g., shaft, bolt, nut, washer, or keyway). Having chosen the elements of the assembly that you wish to present in cross-section, define the section cut depth, i.e., the contour up to which you would like to cut (Fig. 14.12).

For the cross-section area you can use auto hatching (***Auto hatching***). By using auto hatching you deliberately decide for the hatching of individual components to depend on the choice of material and default hatching for a particular material. When you would like to change the hatching for a particular part (a bearing, for example)

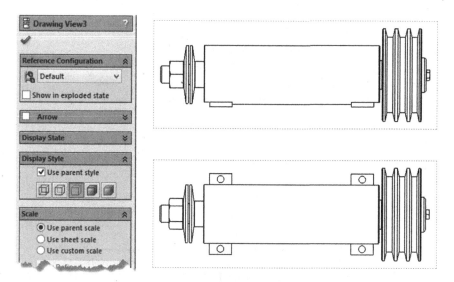

Fig. 14.11 Inserting a projection view in the drawing (note that *centrelines* are still not marked)

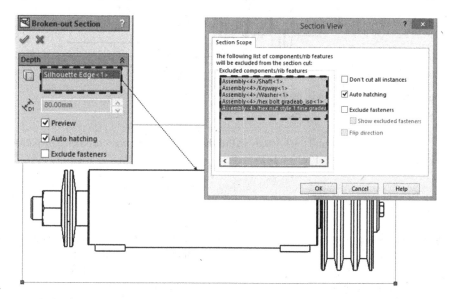

Fig. 14.12 Creating broken-out section across the whole front view

you can apply the change directly and specifically for the cross-section area. Select the area or the component for which you would like to change the hatching. In the properties menu a window will open where you can switch off the auto hatching for a particular material (***Material crosshatch***). Select hatching (***Hatch***) and in the window select a type of hatching, scale and orientation (Fig. 14.13). This will change the hatching for the entire model.

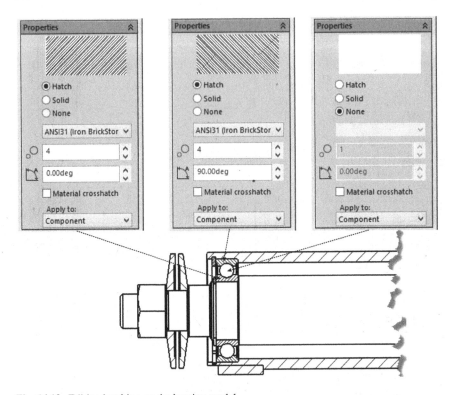

Fig. 14.13 Editing hatching on the bearing model

Positioning the bolt and the keyway in the shaft assembly is presented on a broken-out section of the shaft (Fig. 14.14).

14.3.2.4 Section View

To add a cross-section onto a drawing, activate the (***Insert > Drawing View > Section***) cross-section creation command. A window will open in the properties menu, where you can select a type of cross-section and the section line. This is followed by positioning in the graphic window of the section line to the existing view. Confirm the selected activity and position the cross-section on the drawing (Fig. 14.15).

14.3.2.5 Detail View

A detail view (***Detail View***) in a drawing shows a detail of a model or a magnified view of a marked part. In order to create a detail view, activate the (***Insert > Drawing View > Detail***) detail view creation command. Draw a sketch (circle) of the part close

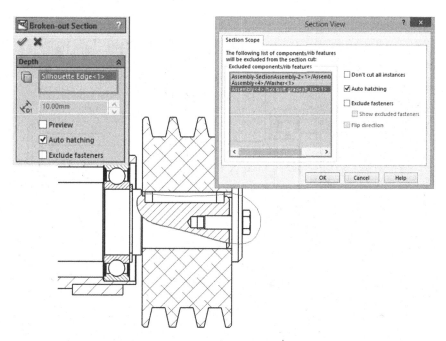

Fig. 14.14 Broken-out section of the shaft, presenting the position of a keyway and a bolt

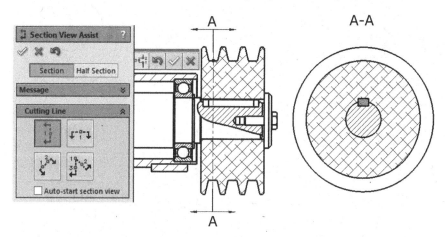

Fig. 14.15 Creating the cross-section on the drawing

to a product's detail (Fig. 14.16). Mark this circle-marked detail with a capital letter and position it in a visible place, next to the marked detail. Never write a marking letter into the hatched part of the cross-section or in the vicinity of the auxiliary dimensioning lines. Poorly marked details or cross-sections in a drawing are proof that the engineer, who does not put this mark in a visible place, does not understand the importance of clear technical documentation (Fig. 14.16).

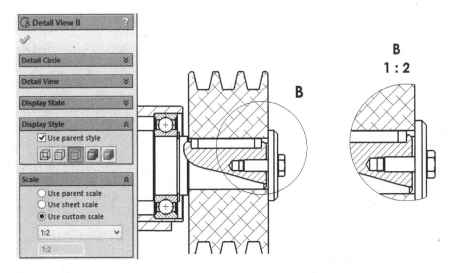

Fig. 14.16 Creating a detail view in the drawing

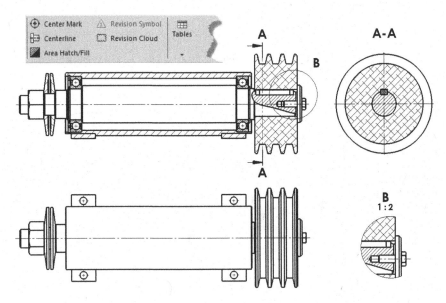

Fig. 14.17 Completing the views in the figure with *centrelines* and *centre marks*

When all the views have been marked with the scale of magnification and cross-sections (including potential details), you can begin equipping the plan. Some of the important supplements include centrelines (*Centerline*) and centre marks (*Center Mark*) (Fig. 14.17).

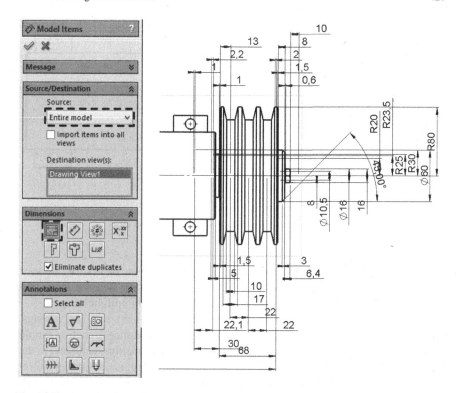

Fig. 14.18 Automated transfer of dimensions from a model into a drawing (view). A group of dimensions is displayed; however, they are not technological and sensible for dimension control

14.3.3 Dimensioning

The dimensioning of a product on a plan should be directly linked to the real model's dimensions, and each change in value should be reflected in a changed shape of the product. All the dimensions are usually given during the stage of creating a model, which makes it possible to import them directly to the drawing or a particular view (*Insert > Model Items …*). When transferring the dimensions, specify a source (an entire model, assembly, individual features, etc.) and select the elements that you wish to insert into the drawing (Fig. 14.18). Such an automated approach usually displays all of the model's dimensions. Such a drawing becomes unclear. Editing these dimensions is generally more time consuming than the direct entry of the dimensions that you wish to present in a particular view and that are in accordance with the technologicality of displaying dimensions and the control concept.

Presented below is direct dimensioning, which will explain the crucial difference between automated and logical technological dimensioning.

Dimensioning can be executed with the *Smart Dimension* (*Tools > Dimensions > Smart*) command. This is followed by selecting the entities that you wish to

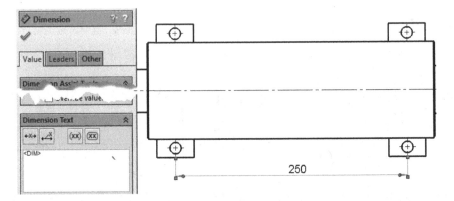

Fig. 14.19 Dimensioning the wheelbase (the dimension still lacks tolerance)

dimension, and positioning the dimensions in the drawing (Fig. 14.19). The value of the dimensions is taken directly from the model's dimensions. Because the model was designed in millimetres, the dimensions are also given in millimetres.

A plan acquires a special value with the insertion of tolerances and matings. Create them by dimensioning the distance for which the tolerance or mating should be created, followed by specifying the parameters in the properties menu and in the *Tolerance/Precision* window. For the mating between the shaft and the pulley, select the (*Fit*) mating and determine the type of mating, i.e., transitional (*Transitional*) mating. Only then do you select the hole and shaft tolerances. Create the deviation table as a new table with the (*Insert > Table > General Table …*) command. Because the data are related, you can merge two cells and insert a measure. Enter the deviations into the cell on the right (Fig. 14.20).

14.3.4 Marks in the Figure

Supplementing marks in a drawing (notes, surface finish, geometric tolerances, a bill of material, position numbers, etc.) add a special value to a manufacturing plan. All these extra features can be added to a drawing with the *Annotation* (*Insert > Annotations > etc.*) command. Specific marks are, of course, typical of specific types of drawing. More characteristic data for different drawing purposes should be added separately onto the drawing plane.

When creating an assembly drawing, consider also adding a bill of material and position numbers as a special feature.

In the case of a manufacturing drawing, the surface finish and the geometric tolerances are added. For welded pieces, add weld symbols and a cut list. Drawings of sheet metal products should include the bending marks etc.

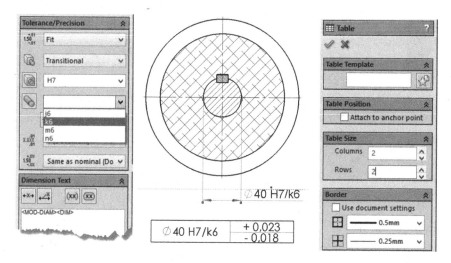

Fig. 14.20 Dimensioning the mating between the shaft and the pulley with a corresponding deviation table. Particular tolerance as an H7 or k6 are define at programme, mating should be calculated by the designer

All these features represent a special set of commands, requiring the pre-preparation of shapes, marks and additional menus. For larger companies, this is carried out together with installing the programmes; however, it would be fair to make this kind of pre-setting generally available and, first of all, useful.

14.3.4.1 Bill of Materials

With a bill of materials being a key piece of information, SolidWorks creates it automatically. Prior to its automated creation, you should define the parameters you wish to include in the bill of materials. General parameter names and then each separate part should be defined at the level of the part or a set. It is vital for the data to be available whenever this part (sub-set or set) is used. It makes it possible for each element of the assembly to be assigned properties that are then displayed in the title block.

With the automated data transfer turned on, it is necessary to open each individual element of the assembly (e.g., *Shaft.sldprt*) and define its properties (*File > Properties, …*). In the (*Properties*) area, enter all the parameters that provide a comprehensive description of a single part (Fig. 14.21).

After entering the data into a special window, representing the bill of material, the latter can be transferred onto the drawing (*Insert > Tables > Bill of Material, etc.*). During the transfer, select a view and specify the bill of material's parameters. Perform the positioning of the bill of materials on the drawing, according to the procedure shown in Fig. 14.22.

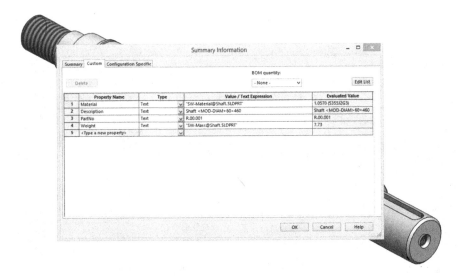

Fig. 14.21 Defining the properties of the shaft model at parameter entry

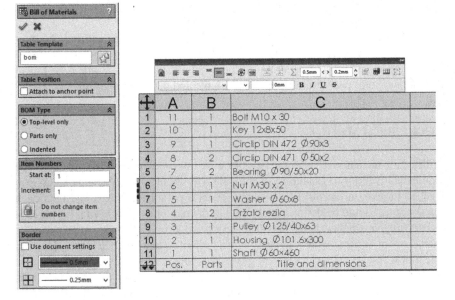

Fig. 14.22 Inserting the bill of materials onto the drawing

14.3.4.2 Position Number (Balloons)

Balloons represent a link between the bill of materials and the model. They determine the position of each element on the plan in the assembly. SolidWorks provides

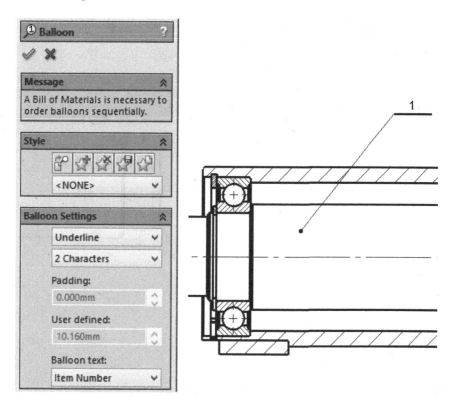

Fig. 14.23 Manual insertion of balloons onto the drawing

the automated creation of balloons (***Auto Balloon***). You need to pay attention to the shape, required by the standard in your *country*. In this case it is also better to use a direct method for specifying the balloons with the (***Balloon***) command. The positioning itself is performed with the (***Insert > Annotations > Balloon, …***) command. It sets the properties and selects the element to be marked, followed by positioning the symbol in the drawing (Fig. 14.23).

14.3.4.3 Surface Finish

The surface condition or, put simply, surface finish, is prescribed by appropriate symbols in the drawing. The (Surface Finish) symbol is added by activating the (***Insert > Annotations > Surface Finish Symbol, etc.***) command, followed by specifying the parameters and positioning the symbol on the drawing (Fig. 14.24).

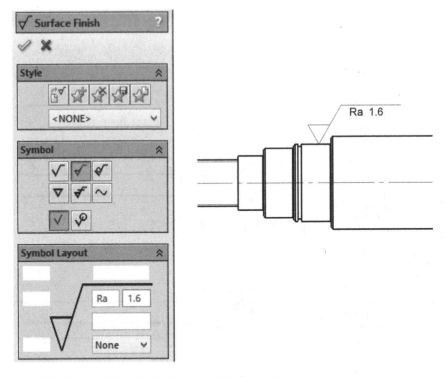

Fig. 14.24 Surface finish in the bearing, treated by fine turning

14.3.4.4 Weld Symbols

The welds in the drawings can be represented by symbols (Weld Symbol). They are added with the (*Insert* > *Annotations* > *Weld Symbol, …*) command. It specifies the type and the size of the welds and other parameters. The symbol has to be positioned on the drawing (Fig. 14.25).

14.3.5 The Title Block of the Drawing

The title block of the drawing or the plan represents an important piece of information for the entire drawing, which makes its editing vital for the information recognition of the document. Editing the title block is performed by right-clicking on the drawing and selecting the *Edit Sheet format* (*Edit* > *Sheet Format*) command. The title-block editor or the title-block template will open. You can complete the information in the title block manually or you can use the automated title-block completion. In this case it makes sense to use the automated data transfer, which minimises any possible mistakes.

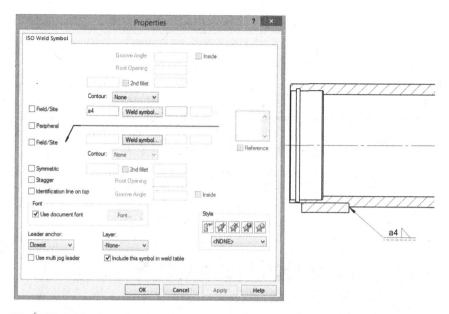

Fig. 14.25 Adding weld symbols to specific and—most importantly—clear spots

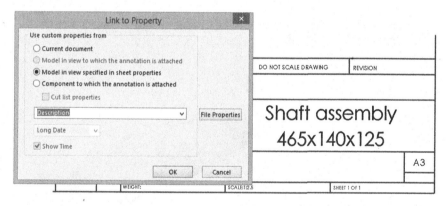

Fig. 14.26 Adding a link for the title in the title block with a model property

With automated completion, the contents of the fields should be linked up with model properties, shown in the figure. Figure 14.26 shows the creation of a link between the title in the title block with the model, presented in the drawing. Begin by selecting the field where you wish to add the links and then select the *Link to Property* option.

Save the edited and prepared title block of the drawing as a sheet format (*File > Save Sheet Format …*). Having entered all the new or linked data, use the (*Edit > Sheet*) command to finish the data entry.

14.4 Modelling in NX

There are two options to create 2D drawings.

1. Begin a new drawing with the (*File > New > Drawing*) command and choose a format.
2. Continue in the existing file (data file **.prt* is the same for the drawing and the piece), and then switch to the drawing module (*Application > Drafting*) and begin importing a previously created 3D assembly model (Fig. 14.27).

It should be noted that it is also possible to change the settings in connection with the drawing standards in the settings (*File > Preferences > Drafting, Utilities > Customer Defaults*). Select a sheet format in the (*Sheet*) window. Under this window, you can choose some fast sheet-format settings (*Edit Sheet*).

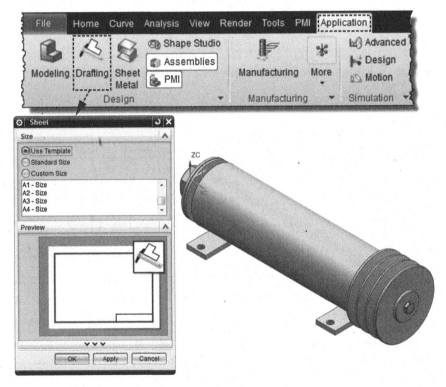

Fig. 14.27 Switching to a module that uses standard formats and plan descriptions, used to create drawings by presenting a 3D assembly model

select sheet layer

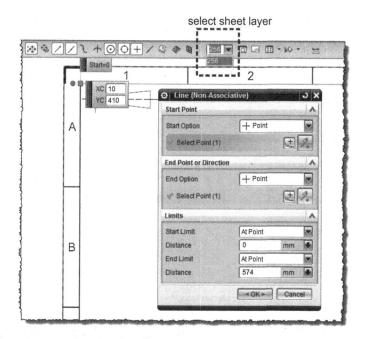

Fig. 14.28 Modifications on the existing title block of the drawing by selecting a sheet layer (*Layer*)

14.4.1 Format and the Title Block of the Drawing

Due to the specific format of technical documents, the drawing environment comes with options that are characteristic of the plans. A window opens in this environment to choose the format and various drawing templates. Similar to most modern modellers, some of the drawing templates are already included in the programme; however, they often do not comply with ISO and other standards. If some other template layout is required, you can create it and include in the programme. By doing so, you can subsequently make the selection from the start menu. But there is also an option to simply modify the existing templates (Fig. 14.28).

In connection with designing the title block of the drawing, you can reset some of the parameters using the ***Drafting Tools > Drawing format*** command (Fig. 14.29). To complete the empty spaces in the title block of the plan, use the (***Populate Title Block***) command. The ***Define Title Block*** command defines a table layout as well as the title block of the drawing. By activating the ***Borders and Zones*** command, it is possible to specify the border of the drawing. Such modified shapes and contents in the current file can be changed into a new template with the ***Mark as Template*** command.

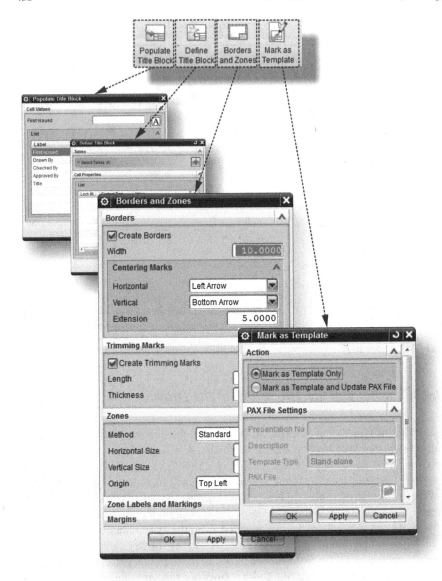

Fig. 14.29 Tools for designing the title block and preparing the templates *Drafting Tools > Drawing format*

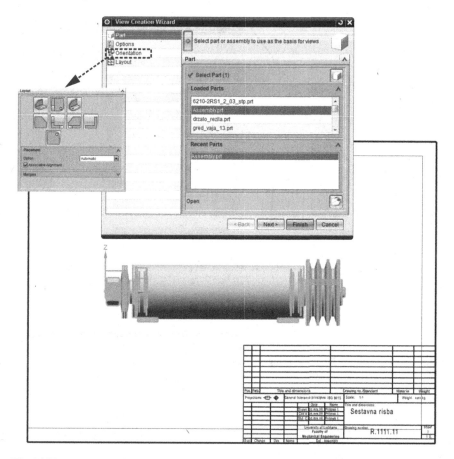

Fig. 14.30 Importing a 3D model using the wizard (*View Creation Wizard*)

14.4.2 Defining Views of 3D Objects

You can transfer views onto the drawing field using the (*View Creation Wizard*) wizard. With the wizard you can choose from a number of orthogonal projections whose orientations and layouts should be set in advance (Fig. 14.30).

3D model import using the base view (Fig. 14.31). Figure 14.32 shows a symbol of the coordinate system, defined according to the selected view. This mark provides good control for any creator of documentation; however, it becomes disturbing later when the drawing is being completed. You can hide the coordinate system mark by right-clicking and selecting *Hide*. If there are no other requirements, you can proceed with your activities and creat a broken-out cross-section (*Section View*). In most cases of presenting the details and the inside, this is vital for a high-quality presentation of the model.

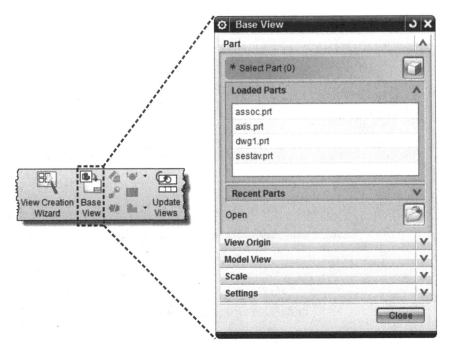

Fig. 14.31   Importing a 3D model using the base view (*Base View*)

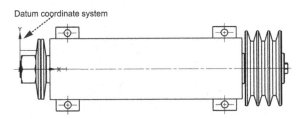

Fig. 14.32   Setting the base orthogonal assembly projection

A further presentation of the assembly's details often includes 3D model entities, such as auxiliary plane sketches, etc. (**Entire Part**). Before that you need to right-click on a component in the *Assembly Navigator* tree structure and select its display using the *Replace Reference Set* command. With this command, you can also include 3D model entities (**Entire Part**) such as auxiliary plane sketches, etc. Selecting *Replace Reference Set > Model* will display only the body (Fig. 14.33).

When creating the section views, the axisymmetric and elongated elements should not be cut. It significantly improves the clarity of the display and you should strictly follow it in the technical documentation. This makes the section view procedure vital, which is why the procedure and the location of the section views should be set in advance, sometimes with an auxiliary sketch on paper or a computer. First, select the

Fig. 14.33 Displaying the auxiliary geometric model entities in a drawing (*Replace Reference Set*)

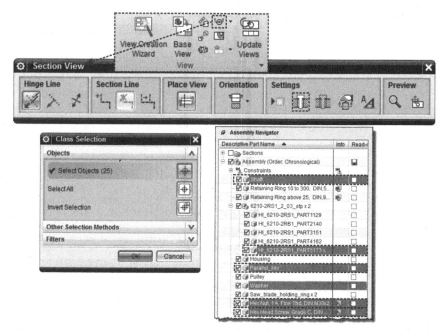

Fig. 14.34 Setting section views and excluding some of the components

view to section, which is followed by placing the section view in the desired position (*Place View*). Continue by setting the view orientation and excluding the components (*Settings*) in the tree structure *Assembly Navigator* that you do not wish to section (Fig. 14.34). The hatching of the sectioned parts is automatic, in accordance with the user's settings (*Customer Defaults*). Examples for sections are defined on Fig. 14.35.

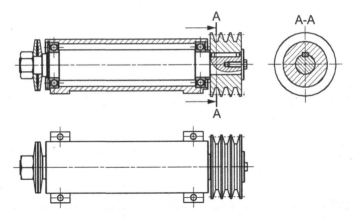

Fig. 14.35 Examples of finished sections A–A

Fig. 14.36 Changing the
hatch pattern on the sectioned
parts

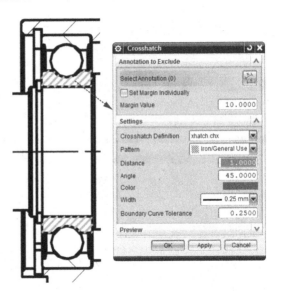

When you wish to change the orientation and the size of the hatching on the sectioned parts you can do this by right-clicking the location of the sectioned and already hatched part and changing the setting within the **Crosshatch** command (Fig. 14.36).

Having checked a drawing or a plan, you might need to correct some details in the drawing. Correcting the details on a drawing in the selected views can be performed with a sketch, like in 3D space. The procedure is as follows: choose a view, use the (**2D Layout > Sketch**) command and make any necessary corrections. Also in this case, you can use the option to hide the finished or—to put it better—marked entities by right-clicking a marked entity and using the **Hide** command (Fig. 14.37). There is

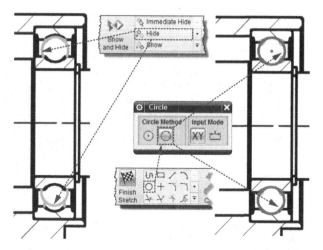

Fig. 14.37 Correcting a drawing on the example of bearing balls by means of a sketch in a view

Fig. 14.38 Sketch contained in the view (**a**) and sketch contained on the drawing sheet (**b**)

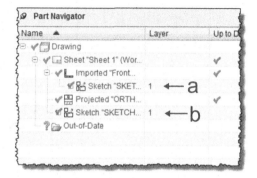

of course the **Show** command that makes it possible to have a group or an individual entity visible again and to show it again.

The location of the sketches in the structure that provides assistance in the drawing is very important (Fig. 14.38). In the event that you use a feature, tied to a certain view, and it is needed for this particular sketch, you have to make sure that this sketch is contained in the view. That process is used if we have more pictures inside one drawing. This is done by right-clicking on the selected view and choosing the (**Activate Sketch View**) command.

14.4.3 Break-Out View

When making a section view, you should roughly take into account the following three steps (Fig. 14.39):

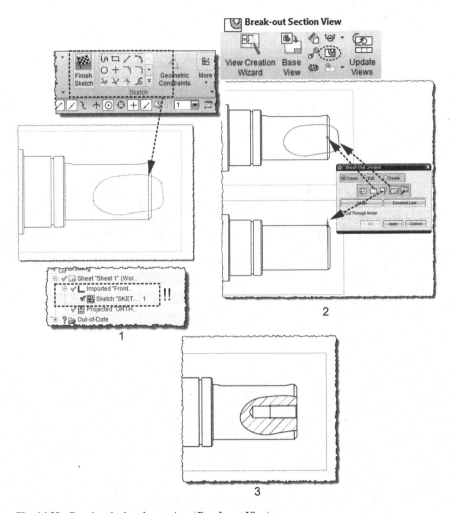

Fig. 14.39 Creating the break-out view (*Break-out View*)

- First, create a sketch of the break-out (1) and finish it at the end.
- Choose the ***Break-out Section View*** command and select the starting position of the break-out and the viewing vector. At the end, select the contour of the sketch that was created earlier in the view (2).
- Finish the procedure and set the hatching pattern if necessary (3).

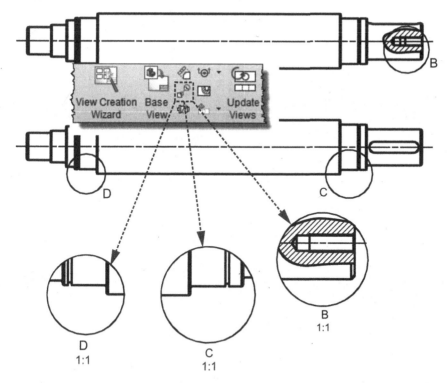

Fig. 14.40 Creating detail views (*Detail View*)

14.4.4 Creating Details (Detail View)

To create detail views, use the (*Detail View*) command. Usually, you need to define a viewing type (circular, rectangular and size ratio (*Scale*). Detail views can usually be freely moved across the sheet. Each dimension, tied to the view (*View*), moves with it (Figs. 14.40, 14.41).

14.4.5 Dimensioning

To present the measuring information in the drawing, use the commands from the *Dimension* set. This feature provides an additional set of commands to determine the length, angles and other dimensions (Fig. 14.42). The most frequently used one is *Rapid Dimension*, where the programme automatically chooses the dimensioning method according to what the user selects with a click.

Fig. 14.41 Detail view (*Detail View*) feature settings include several options for viewing types and detail layout

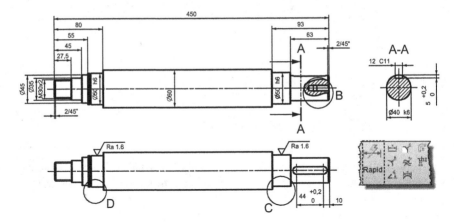

Fig. 14.42 The created dimensions are tied to particular views and follow them with the change of location

However, you should be careful with selecting the option for automated dimensioning. Namely, due to the specific shapes of new products, it is possible that a technologically misleading method for defining measures and dimensions might appear.

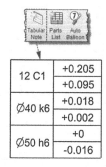

12 C1	+0.205
	+0.095
Ø40 k6	+0.018
	+0.002
Ø50 h6	+0
	-0.016

Fig. 14.43 Creating tolerance and mating tables (*Tabular Note*)

		Projection	Edges: ISO 128		General tolerances:	Surface:	Scale: 1:2 (1:1)	Pos.: 1	Weight: 7,92 kg
			internal	external		ISO 1302	Material:		
			⊟← ⊕↓		ISO 2678-mK		St 37.2		
12 C1	+0.205		⌊₀,₂	⌊₀,₃					
	+0.095				Date	Name	Title and dimensions:		
					Drawn dd.mm.y	Zorko D.			
Ø40 k6	+0.018				Chk'd dd.mm.y	Demsar	Shaft Ø 60x450		
	+0.002				Std. C. dd.mm.y	Duhovnik			
	+0				University of Ljubljana		Drawing number:		Sheet
Ø50 h6					Faculty of		R.1111.11		1
	-0.016	Sign	Change	Day	Mechanical Engineering				1 S.
					Name Basic drawing:		Replaced:	Replaced w.:	

Fig. 14.44 Attaching the mating table to the title block of the assembly drawing

14.4.6 Tolerances and Matings

Tolerances and matings are presented in special tables (Fig. 14.43) (*Table > Tabular Note*). Determine a table layout or choose it according to the number of columns and rows. The cells can also be merged. When a table has been created and equipped with data you can simply attach it to the title block of the drawing (Figs. 14.43, 14.44).

14.4.7 Miscellaneous Marks in the Drawing

14.4.7.1 Creating a Bill of Material

A bill of material is the most important information to link an assembly with manufacturing plans and a purchase parts list. It should be precisely and clearly created. In principle, this is a table containing information about the assembly components, such as name, serial number, mass, etc. Information about the components can be automatically imported into the table on the drawing under the condition that everything had been set and inserted in an appropriate memory field, separately for each component (Fig. 14.45).

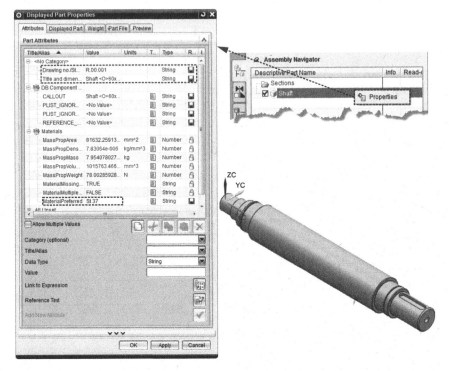

Fig. 14.45 Setting component data, used for the bill of material

Activate the bill of material with the command in the table set (*Table > Parts List*). Table creation commands can also be directly reset by changing the text in the cells and/or by adding or removing the lines and columns (*Tabular Note*) (Fig. 14.46).

14.4.7.2 Marking Positions (Balloon)

The positions, specifying the place and location, can be left to the programme to execute the marking automatically with the *Auto Balloon* command. When a suggested form of the position does not conform to a particular standard, the positions and form can be set directly (*Balloon*) (Fig. 14.47).

14.4.7.3 Marking Surface Finish

Activate surface finish with the *Surface Finish* feature, where all the necessary settings are available to mark the surface quality (Fig. 14.48).

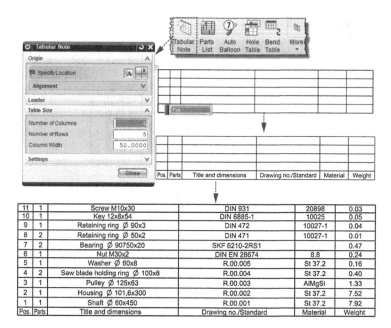

Pos.	Parts	Title and dimensions	Drawing no./Standard	Material	Weight
11	1	Screw M10x30	DIN 931	20898	0.03
10	1	Key 12x8x54	DIN 6885-1	10025	0.05
9	1	Retaining ring ⌀ 90x3	DIN 472	10027-1	0.04
8	2	Retaining ring ⌀ 50x2	DIN 471	10027-1	0.01
7	2	Bearing ⌀ 90750x20	SKF 6210-2RS1		0.47
6	1	Nut M30x2	DIN EN 28674	8.8	0.24
5	1	Washer ⌀ 60x8	R.00.005	St 37.2	0.16
4	2	Saw blade holding ring ⌀ 100x8	R.00.004	St 37.2	0.40
3	1	Pulley ⌀ 125x63	R.00.003	AlMgSi	1.33
2	1	Housing ⌀ 101,6x300	R.00.002	St 37.2	7.52
1	1	Shaft ⌀ 60x450	R.00.001	St 37.2	7.92
Pos.	Parts	Title and dimensions	Drawing no./Standard	Material	Weight

Fig. 14.46 Direct completion of the bill of material and its final layout

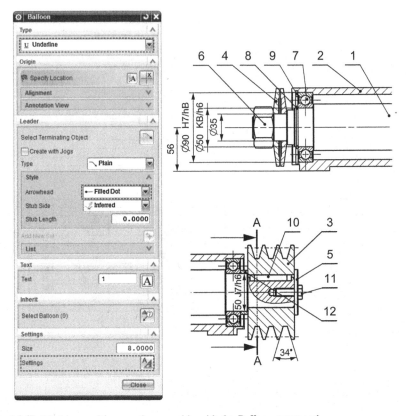

Fig. 14.47 Marking positions on the assembly with the **Balloon** command

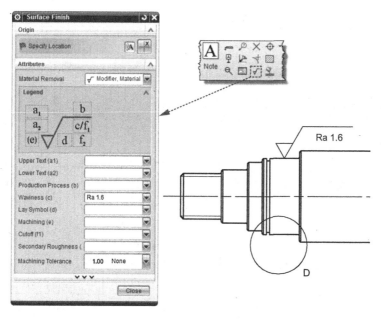

Fig. 14.48 Marking the piece's surface finish on the drawing (*Surface Finish*)

Fig. 14.49 Writing text on the drawing with the *Note* command

Notes:
All undimensioned chamfers are 1/45°
All undimensioned roundings are R=0.5 mm

14.4.7.4 Comments on the Drawing

The *Note* command allows adding comments to the drawing at a selected location (Fig. 14.49).

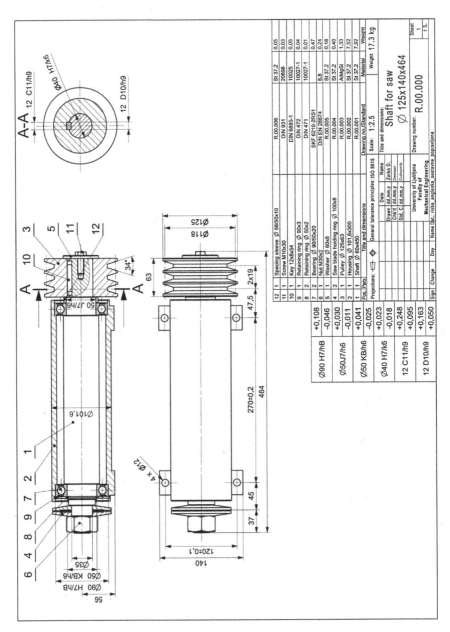

Fig. 14.50 Saw shaft (complex) 125 × 140 × 464 mm

14.5 Examples

Figures 14.51, 14.52 and 14.53 preset few examples for technical documentation.

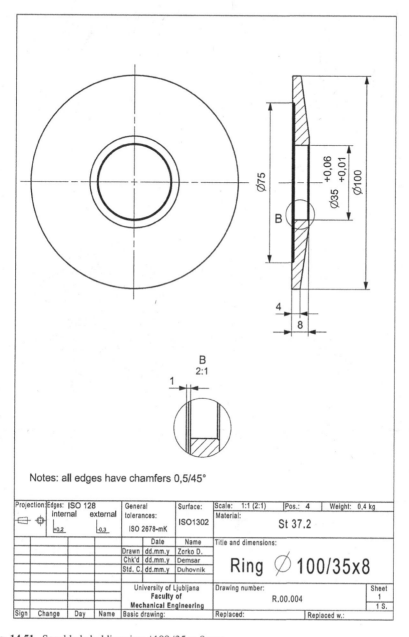

Fig. 14.51 Saw blade holding ring $\phi 100/35 \times 8$ mm

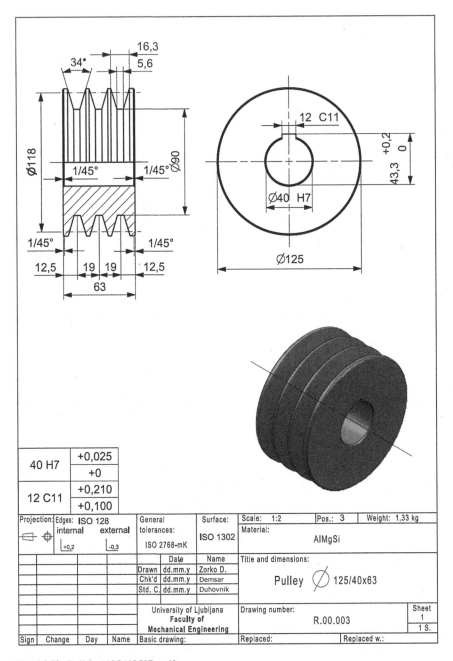

40 H7	+0,025
	+0
12 C11	+0,210
	+0,100

Projection:	Edges: ISO 128		General tolerances:		Surface:	Scale: 1:2		Pos.: 3		Weight: 1,33 kg
⊲⊐ ⊕	internal	external	ISO 2768-mK		ISO 1302	Material:				
	⌊+0,2	⌊-0,3				AlMgSi				
			Date	Name		Title and dimensions:				
			Drawn	dd.mm.y	Zorko D.					
			Chk'd	dd.mm.y	Demsar	Pulley ⌀ 125/40x63				
			Std. C.	dd.mm.y	Duhovnik					
			University of Ljubljana **Faculty of** **Mechanical Engineering**			Drawing number:		R.00.003		Sheet 1
										1 S.
Sign	Change	Day	Name	Basic drawing:		Replaced:			Replaced w.:	

Fig. 14.52 Pulley $\phi 125/40 H7 \times 63$ mm

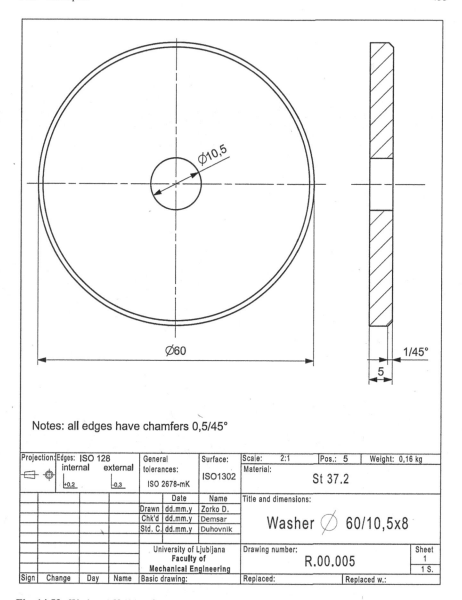

Notes: all edges have chamfers 0,5/45°

Projection:	Edges: ISO 128		General tolerances:	Surface:	Scale:	2:1	Pos.: 5	Weight: 0,16 kg
	internal	external		ISO1302	Material:			
	+0,2	-0,3	ISO 2678-mK		St 37.2			
			Date	Name	Title and dimensions:			
			Drawn dd.mm.y	Zorko D.				
			Chk'd dd.mm.y	Demsar	Washer ⌀ 60/10,5x8			
			Std. C. dd.mm.y	Duhovnik				
			University of Ljubljana Faculty of Mechanical Engineering		Drawing number: R.00.005			Sheet 1 1 S.
Sign	Change	Day	Name	Basic drawing:	Replaced:		Replaced w.:	

Fig. 14.53 Washer $\phi 60/11 \times 8$ mm

Chapter 15
Modellers and Technical Documentation

Abstract Technical documentation includes all the information about a product, with a plan, or a drawing, being its key element. Engineers take a plan or a drawing as the key information, and then the information, based on the plan. This chapter presents the abilities and deviations of making drawings of specific details that significantly qualify the quality of a particular modeller. Due to differences in the development phases there are still some pieces of key information and messages that qualify a plan but they are unfortunately not yet included in the standard software. Should future developments include building international standards of technical documentation into the modellers, it would not only simplify, but most of all raise, the general technical culture. This chapter will present some typical deviations in the existing software, accessible on the market but not taking account of the proper use of ISO standards in its basic version.

15.1 The Size of Written Text and Dimensions

The recommended contour thicknesses for the parts of an image, a plan or a drawing are shown in Table 15.1. The recommendation is based exclusively on the clarity and explicitness of the object on the drawing plane, particularly the proportion between the perception of the line thickness with the naked eye and a broad view of the whole drawing plane. The general condition is a one-min arc resolution, which is logical and essential for all graphical representations. At a 1-mm line thickness, one min represents at least 0.30 mm. According to the Table 15.1, a suitable thickness is at least 0.50 mm. It should be noted that the format size is A 00 841 × 1194 mm. A format of this size is looked at from a distance of at least between 0.8 and 1.0 m. Of course, the size of the letters and the dimensions follows the format. Due to the small size of screens, beginners often encounter problems with computer-aided designing because they do not see a distinction between the virtual world and real dimensions.

J. Duhovnik et al., *Space Modeling with SolidWorks and NX*,
DOI: 10.1007/978-3-319-03862-9_15, © Springer International Publishing Switzerland 2015

Table 15.1 Correlation between the size of a drawing, a plan and a line thickness

Drawing area size /	Contour thickness (mm)	Height of written dimensions (mm)	Height of letters for inserted notes (mm)	Height of letters for cross-section and details (mm)	Height of written scale annotations (mm)
A 8 to A 7	0.25	2	3	5	3
A 5 to A 6	0.35	3	3	5	3
A 3 to A 4	0.5	3	4	6	4
A 1 to A 2	0.7	3.5	4	6	4
A 00 to A 0	1.0	4	5	7–8	5
over A 00	1.4 (2.0)	4	5	10	6

The same table includes the sizes of written dimensions and letters for labelling cross-sections, details and scales.

The purpose of suggesting these values lies in the fact that creating technical documentation and testing young engineers has revealed confusion among engineers when looking at plan printouts for the first time. This confusion is even greater in the field when you need to interpret these details. In any case, using a laptop computer is recommended; however, it is vital to have screens exceeding 19 or 20 inches. Applying ratios in a technical drawing provides clarity and, most importantly, the rapid transfer of information to the users.

More complex products are usually represented in cross-sections, executed on planes. They can be either aligned or broken. The representation of cross-sections as solid materials is executed by labelling the sectioned surfaces as solid surfaces, and then hatching them. It is important for the hatching to match the size of the cross-section. A beginner often faces the question of how thick the hatching should be to properly represent a surface. Line thickness is specified by the ISO 128-22 and 128-50 standards.

The line thickness is labelled as shown in Table 15.2. Modellers apply the general thickness, which presents the designer with the question as to what line thicknesses to use to draw a contour, hatching, centrelines, etc. (Tables 15.2 and 15.3).

15.2 Automated Dimensioning

When a beginner—having finished modelling a product—starts making technical documentation, he or she would like to present some of the shapes as quickly as possible, with the standards, defining the method and approach to the representation. With one of the first things being dimensioning, one would expect the software to provide automated dimensioning. There have been some research attempts from the very beginning to introduce automated dimensioning. However, it soon became clear that using or claiming to be using automated dimensioning is completely false. The

Table 15.2 Line thicknesses and their intended use, and the thickness gradation of the lines

No.	Line — Description and representation	Aplication	Reference to ISO
01.1	Continuous narrow line	.1 imaginary lines of intersection	-
		.2 dimesnsion lines	129
		.3 extension lines	129
		.4 leader lines and reference lines	128-22
		.5 hatching	128-50
		.6 outlines of revolved sections	128-40
		.7 short centre lines	-
		.8 root of screw threads	6410-1
		.9 origin and terminations of dimension lines	129
		.10 diagonals for the indication of flat surfaces	-
		.11 bending lines on blanks and processed parts	-
		.12 framing of details	-
		.13 indication indication of repetitive details	-
		.14 interpretation lines of tapered features	3040
		.15 location of laminations	-
		.16 projection lines	-
		.17 grid lines	-
	Continuous narrow freehand line	.18 preferably manually represented termination of partial or interrupted views, cuts and sections, if the limit is not a line of symmetry or a centre line*	-
	Continuous narrow line with zigzags	.19 mechanically represented termination of partial or interrupted views, cuts and sections, if the limit is not a line of symmetry or a centre line*	-

No.	Line — Description and representation	Aplication	Reference to ISO
01.2	Continous wide line	.1 visible edges	128-30
		.2 visible outlines	128-30
		.3 crests of screw threads	6410-1
		.4 limit of length of full depth thread	6410-1
		.5 main representations in diagrams, maps, flow charts	-
		.6 system lines (structural metal engineering)	5261
		.7 parting lines of moulds in views	10135
		.8 lines of cuts and section arrows	128-40
02.1	Dashed narrow line	.1 hidden edges	128-30
		.2 hidden outlines	128-30
02.2	Dashed wide line	.1 indication of permissible areas of surface treatment, e.g. heat treatment	-
04.1	Long-dashed dotted narrow line	.1 centre lines	-
		.2 lines of symmetry	-
		.3 pitch circle of gears	2203
		.4 pitch circle of holes	-
04.2	Long-dashed dotted wide line	.1 indication of (limited) required areas of surface treatment, e.g. heat treatment	-
		.2 position of cutting planes	128-40
05.1	Long-dashed double-dotted narrow line	.1 outlines of adjacent parts	-
		.2 extreme positions of movable parts	-
		.3 centroidal lines	-
		.4 initial outlines prior to forming	-
		.5 parts situated in front of a cutting plane	-
		.6 outlines of alternative executions	-
		.7 outlines of the finished part within blanks	10135
		.8 framing of particular fields/areas	-
		.9 projected tolerance zone	10578

* It is recommended to use only one type of line on one drawing

Table 15.3 Standardized line thicknesses and the ratios between the lines according to their intended use (ISO 128-20 to 128-50)

	Basic line thickness, based on the thickest A (contour) mm								
	0.13	0.18	0.25	0.35	0.5	0.7	1.0	1.4	2.0
A			○	⬭	☆	☺	△	⬭	○
B	○	⬭	☆	☺	△	⬭	○		
C		○	⬭	☆	☺	△	⬭	○	
D		○	⬭	☆	☺	△	⬭	○	
E			○	⬭	☆	☺	△	⬭	○
F		○	⬭	☆·	☺	△	⬭	○	
G	○	⬭	☆	☺	△	⬭	○		
H	○	⬭	○☆	⬭☺	☆△	☺⬭	△	⬭	○
J			○	⬭	☆	☺	△	⬭	○
K	○	⬭	☆	☺	△	⬭	○		

Legende:

○   extreme range

⬭   thickness line

☆   used standard format

☺   used bigger format

△   used by biggest format

⬭   wide line range

○   extreme width range

dimensioning paradigm requires taking account of a pre-chosen technology or even a specific technological operation (Figs. 15.1, 15.2 and 15.3).

The argument that it is possible to turn automated dimensioning on in order to obtain the characteristics of specifying particular shapes, without regard to a particular technology, is false and misleading for beginners. All modellers usually show all the dimensions in a drawing, specified during the model's creation. In order to be able to represent a model clearly and at the same time taking account of the manufacturing and control technologies, manual dimensioning is the most suitable method. For the purpose of understanding the differences between the automated and proper dimensioning of certain shapes, some examples are presented below, beginning with the automated and proper dimensioning of a part, respectively (Figs. 15.4, 15.5, 15.6, 15.7 and 15.8).

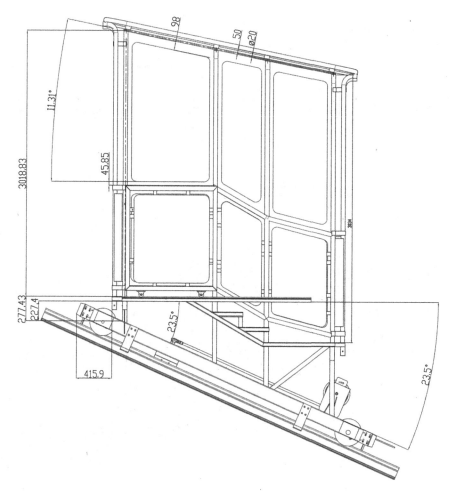

Fig. 15.1 Trolley for passengers presented with the same thickness of line. A common mistake in the presentation of designers who used poor capabilities of CAD modellers. (Example: Planica 2014, Slovenia)

Because the examples are clear enough, they are not explained further.

15.3 Labelling Cross-Sections

Labelling cross-sections in each software package should already be defined in the main menu. In an NX modeller, define it in the *Customer Defaults* menu. Detailed settings, defining technical documentation, are adjusted in the *Drafting* sub-menu (Fig. 15.9). In our case, you can set a standard (ISO, BS, AGMA, JIS, etc.) that a

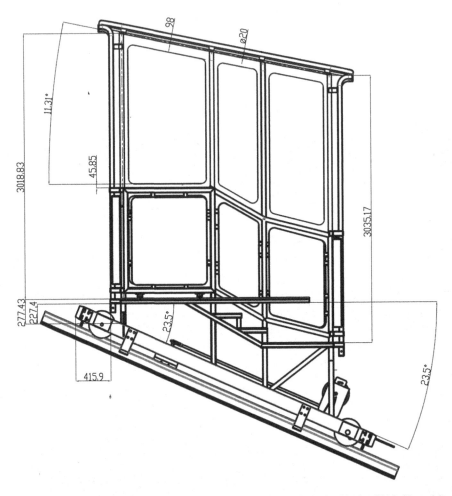

Fig. 15.2 Correctly presented trolley with the contour visible. (Example: Planica 2014, Slovenia)

programme should follow to create the documentation. Individual standards differ from one another in their representation of details and detailed methods of representations. The basics are recognizable and identical, while the differences in representing cross-sections, screws, positioning, etc. are significant. It can pose a major challenge for modeller developers because seemingly simple routines require a logical general treatment if representation is different from country to country. With a view to standardizing representations as much as possible, there are identical standards for larger economies and larger areas. The EU, for example, uses the ISO standard. Due to the large size of the EU and with it representing a significant industrial power, one would expect ISO standard labels to come with the solutions that are properly ISO-standardized. Unfortunately, practice has shown that modellers are not as sophisticated in this respect, which requires a lot of extra settings and manual

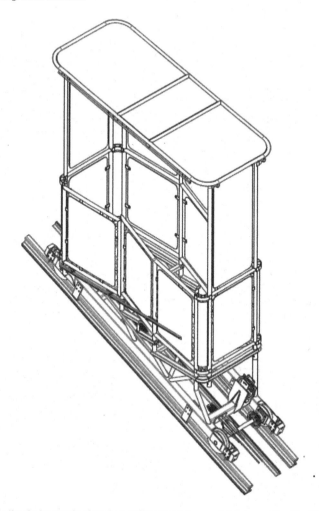

Fig. 15.3 Trolley in isometric view (use different thickness of line according the ISO 128–20 to 128–50)

modifications to plans. It is also a reason for the indirect restriction of access to a wider user environment. It is a fact that engineering work accounts for around 20 % of the engineering work of product development and 70 % is accounted for by technical work, where the larger part consists of complementing technologies and the details of technical documentation. Presented below are some typical examples that need to be adjusted manually by means of additional setting and correcting plans if you are to create a drawing, conforming to the ISO standard.

When representing a product on a plan, the use of cross-sections (whole, half and partial) is a key piece of condensed information. For this reason, we will first present the differences between programmed representations of cross-sections, and

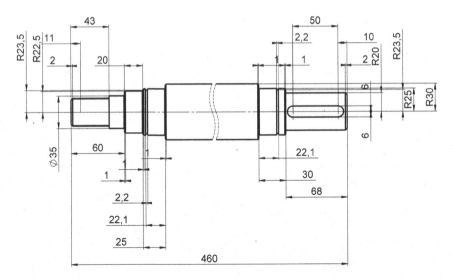

Fig. 15.4 Automated dimensioning of a shaft. The figure shows that a modeller can only take account of the dimensions that are specified during the model's creation. Technology-dependent dimensions are usually different

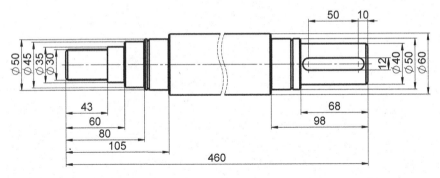

Fig. 15.5 Dimensioning (manual) a shaft according to manufacturing technology (parallel dimensioning), typical of the methods of mounting and knife movements along the lathe

complementing a plan in order to follow the ISO standard. It should be noted that the selection was made in the ISO standard menu. The result of the presented cross-section according to the ISO standard selection is shown in Fig. 15.10. A difference between the executed and actually used ISO standard can be seen by comparing Figs. 15.10 and 15.15. The figures between the main two ones show the necessary steps to achieve manually created conditions of the proper use of the ISO standard.

Regardless of the incorrect representation of a cross-section in a modeller, it is possible—by means of the additional setting of some records—to create an ISO-complying plan. Similarly, corrections can also be made to all other standards. For this purpose, apply the ***Customer Defaults*** additional settings under the

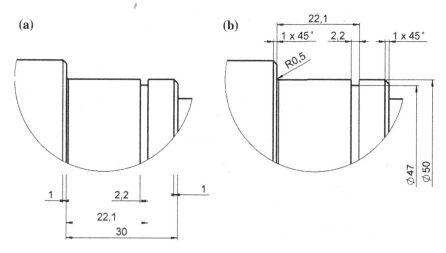

Fig. 15.6 Dimensioning edge blends and chamfering on a detail: automated (a summary of dimensions on the base sketch) (**a**) and manual, taking account of both the manufacturing technology and a functional approach to determining the position of a snap (**b**)

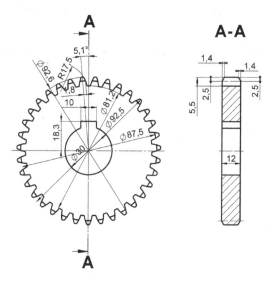

Fig. 15.7 Detailed example of automated dimensioning of a gear, following the principle of inserting the dimensions that determine the shape of the cogwheel

Customize Standard tab. In this sub-menu, you can modify the cross-section labelling. Unfortunately, modifying the settings is confusing and time-consuming. For beginners, this approach means a significant waste of time and they largely avoid it (Fig. 15.11).

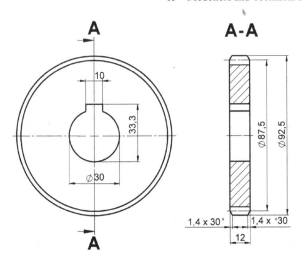

Fig. 15.8 Dimensioning a gear, providing all the key measurements for its manufacture. The drawing is legible and all the dimensions are clearly defined

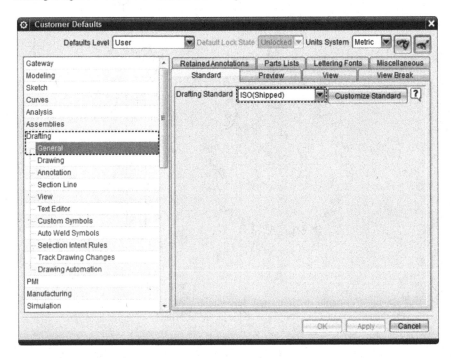

Fig. 15.9 Sub-menu *Drafting*, allowing in *General* selecting *ISO standard* according to which a plan should be made

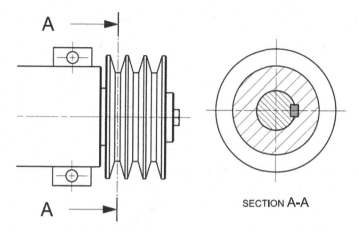

Fig. 15.10 Labelling a section and section plane in the basic settings to define a section on a section plane, in compliance with the ISO standard

Cross-section labelling is modified with the settings that follow the settings in a larger number of tabs. Figure 15.12 shows the settings for cross-section labelling. In the *Position* tab you can set the location of a letter. In our case, the letter should be above the section line. To do this, use the *Above* command. The ISO standard does not require specific section labelling "SECTION", so delete the record with the *Prefix (Section)* command, which results in an empty text box. Later, you will recognize that this is not enough and "SECTION" must be deleted again from the record with a special command. The size of the labels on plans must be proportional. It applies to dimensions, details, cross-sections, annotations, etc. It is the proportion of labels that makes a plan legible, understandable and most of all, useful. Also in our case you need to set a proper size of letters that label a cross-section. This is done with the *Letter Size Factor* command, where you can set the letter size directly. It should be right for the setting to match other text sizes.

The key setting elements to create cross-section labels are executed in the *View Label* sub-menu. Figure 15.11 shows the settings for labelling the section plane on the main view. It should be noted that for *Label Location,* select *On End* in order to move the letters to the end of the line, representing the section plane. Set *Label Distance* to 6 mm because the pre-set value of 12 mm for the particular case moves the letters too far from the label line.

Once everything from above has been set, a cross-section such as the one in Fig. 15.13 will appear. You can see that despite deleting the "SECTION" prefix in the *Prefix* command, it still appears on the plan, although it is redundant according to the ISO standard.

Correcting the "**Section A-A**" label can be performed by double-clicking directly on its location on the screen. Double-clicking opens the setting sub-window, shown in Fig. 15.14. This window allows, in the second step, completing the already chosen settings for representing the cross-section label. With the only problem being the

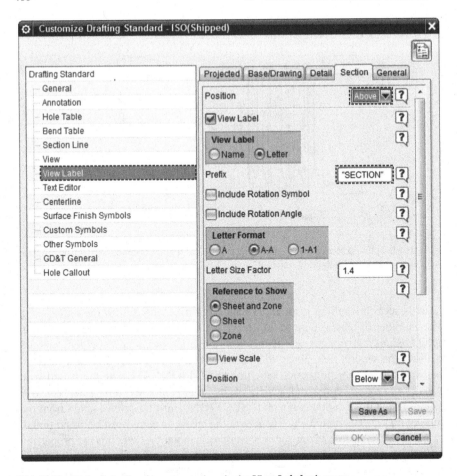

Fig. 15.11 Setting labelling for cross-sections in the *View Label* sub-menu

section label in the Prefix sub-menu, delete "SECTION", and set the letter size with a new letter factor, which is 2 in our case.

After our specific settings, the cross-section is labelled in accordance with the ISO standard, shown in Fig. 15.15. The procedure can be completely automatic if it has been properly introduced in the basic procedure, defined by the *ISO standard* sub-menu.

Finally, we can confirm that the ISO standard is used nearly at all. We do not follow the ISO standard with the arrow, where the line should be type A, not B (Fig. 15.16).

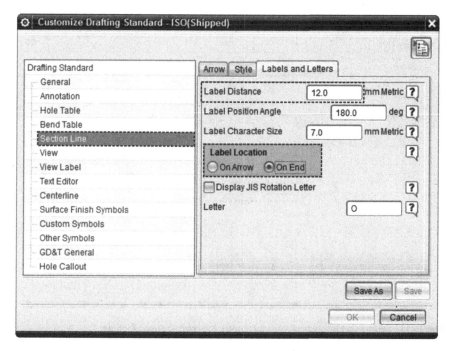

Fig. 15.12 Setting labelling for the section line on the main view

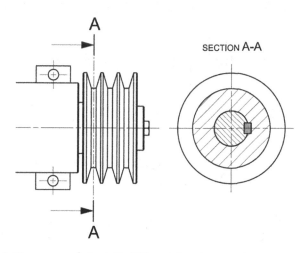

Fig. 15.13 Labelling a cross-section in the NX modeller after changing the settings

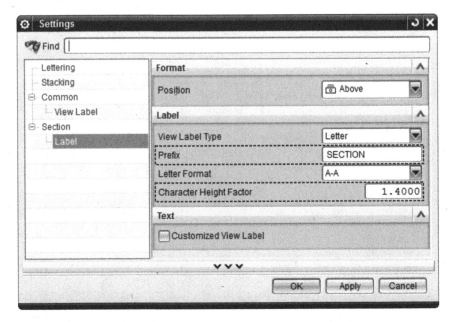

Fig. 15.14   Setting window, where you can modify the cross-section labelling

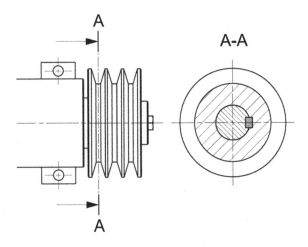

Fig. 15.15   Manual labelling of a cross-section, complying with the ISO standard

15.4 Hatching

Special attention should be paid to the hatching density on the surfaces. The hatching surface is defined by taking account of the longest distance (the longest side), labelled as 'a', and a line, perpendicular to this distance, i.e., the shorter side 'b'. Both characteristic lengths are shown in Fig. 15.17.

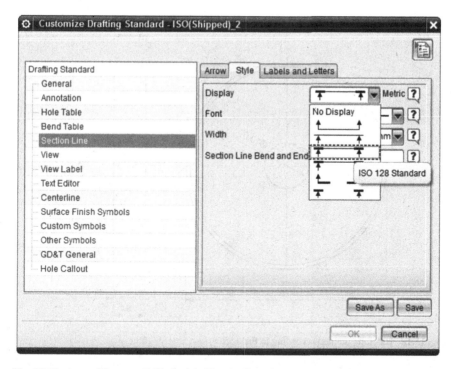

Fig. 15.16 A set of lines, available for labelling the line of the cross-section plane

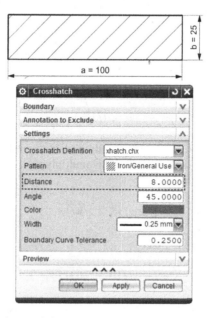

Fig. 15.17 Defining the characteristic dimensions of the section plane. They define the hatching density parameters

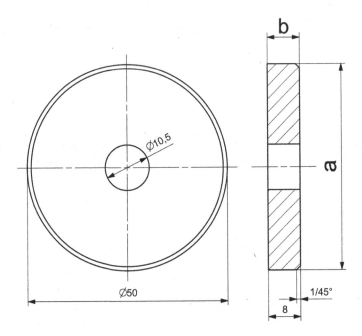

Fig. 15.18 Hatching defined according to Eq. 15.1

The easiest way to determine the distance between the hatching lines is this simple equation:

$$t = a/(10-15) + \sqrt{b/(10-15)} \qquad (15.1)$$

When the distance 'a' exceeds the distance 'b' by more than a factor of 7 (long partial bush cross-sections, shafts along the keyway, etc.), take this simplified equation

$$t = a/15 \qquad (15.2)$$

If the distances 'a' and 'b' are in a relation of between 0.75 and 1.5, also take account of the size of the surface, defined by multiplying $a \times b$.
When $\sqrt{a \times b} < 1{,}000$ mm^2, then

$$t = a/15 + \sqrt{b/15} \qquad (15.3)$$

When $\sqrt{a \times b} > 1{,}000$ mm^2, then

$$t = a/10 + \sqrt{b/15} \qquad (15.4)$$

The mathematical formulae to define the hatching density have been created in order to be able to input them into a software environment and provide the automated specification of a suitable hatching density (Figs. 15.18, 15.19, 15.20 and 15.21).

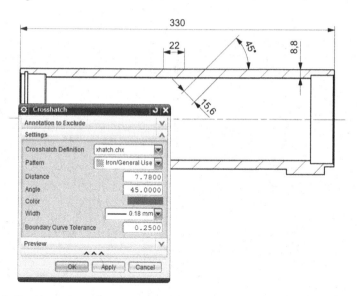

Fig. 15.19 Hatching defined according Eq. 15.2

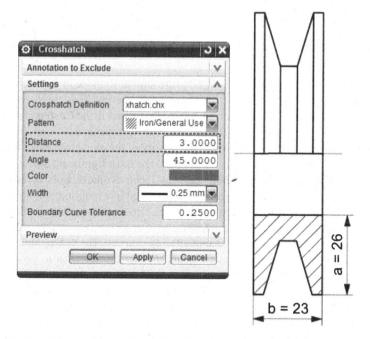

Fig. 15.20 Hatching defined according to Eq. 15.3 using the comparison 'a:b' between 0.70 and 1.5 and small hatching plane, less of 1,000 mm^2 on the drawing

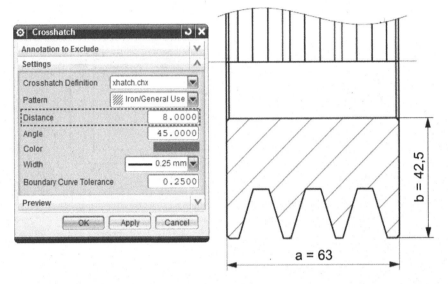

Fig. 15.21 Hatching defined according to Eq. 15.4 using the comparison 'a:b' between 0.70 and 1.5 and small hatching plane, more of 1,000 mm^2 on the drawing

15.5 Presenting the Section Planes of Axes and Screws

The differences in presenting cross-sections are similar to those in their representation. Figure 15.22 shows a cross-section along a shaft bearing. This cross-section was made with the NX modeller after setting the ***ISO standard*** in the ***Customer Defaults*** sub-menu. ISO standard experts will soon spot some significant deviations in the representation, which significantly reduces the quality of the plan, and, most of all, the drawn elements are unrecognizable and unclear. The most significant deviation from the standard is in the case of machine elements, such as shafts, axes, screws, bolts (axial symmetric parts), which are not to be sectioned in a longitudinal direction.

The below follows an attempt to show how to correct a plan to make its clarity and explicitness comply with the ISO standards, and to make a plan useful for a designer, a technologist and all users in the manufacturing process. The NX modellers offer the option of presenting individual elements that we do not wish to present in a cross-section view. The basic point is that a cross-section can be carried out via a 3D model as it is, and the designer chooses the elements that are not to be sectioned according to the standard. The criterion for this representation depends on the size of the diameter, the axis of the element and, of course, its length. As the criteria are rather sophisticated, and the element description is not yet available in the parts list, it is difficult to make the selection of elements automated. If it was required to define the name and dimension of a particular element in order to be able to position it, it would be possible—by means of an additional expert system—to define the elements that are

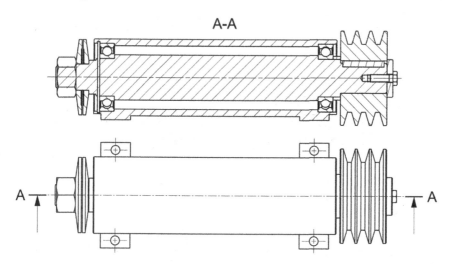

Fig. 15.22 Assembly cross-section, created by a modeller after settings in the sub-menu (*ISO standard*)

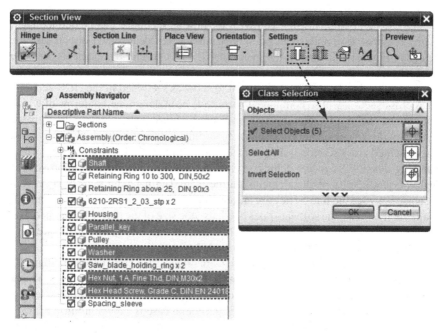

Fig. 15.23 Selecting the elements that are not to be shown as sectioned in the cross-section

Fig. 15.24 Selecting the elements in the assembly to be shown as sectioned in the cross-section

represented as cross-sections or elements. They would be automatically represented in the view.

The choice of elements that you wish to show as sectioned can be defined later, as the required assemblies can be defined only when determined directly by the designer. Figure 15.23 shows the menu that allows defining the elements that you do not want to show as sectioned in the cross-section. Figure 15.24 presents the procedure for the elements that you want to show as sectioned in the cross-section.

A cross-section whose component parts have been determined in terms of which ones should be sectioned and which ones not, is presented in Fig. 15.25. Anyone familiar with the presentation technique according to different standards can see that the presented cross-section is unfortunately still deficient in determining the surface hatching density. Surface hatching of a proper density significantly improves the legibility and clarity of the elements on the plan. The hatching density is set as a distance between the lines. They are automatically pre-set prior to hatching. This is a major fault because the software does not detect the size of the part to be hatched, and the hatching should be corrected manually. This problem is discussed in the previous sub-chapter and a possible solution is provided.

Cross-sections representations on a plan are often very clearly executed, logically formulated and understandable to the reader. In these cases, there is no need to mark the section plane with special labels, nor is it necessary to label the cross-section itself. This representation method for this particular case is shown in Fig. 15.26.

As long as there are only few component parts, selecting the parts to be sectioned poses no major problem. It is also clear where the section plane is and where not. In the event of a larger number of elements and a complex assembly, the section line is usually marked, and the elements to be sectioned are added manually.

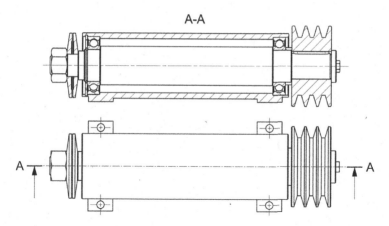

Fig. 15.25 Cross-section executed after specifying which elements of the assembly are to be shown as sectioned and which ones not

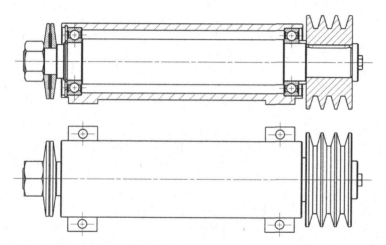

Fig. 15.26 Cross-section, complying with the ISO standard (hatching density and clear cross-sections can be without cross-section labels)

A plastic injection-moulding tool is a good example of the above statement. Labelling a multi-plane cross-section of a simple injection-moulding tool is shown in Fig. 15.27.

The presented plastic injection-moulding tool consists of a large number of standard parts. They can comply with standards or they can be global products, produced in an environment of reliable and quality technology and manufacturing. In these cases CAD models can be imported from the manufacturer's databases. They usually

Fig. 15.27 Multi-plane cross-
section of a simple injection-
moulding tool

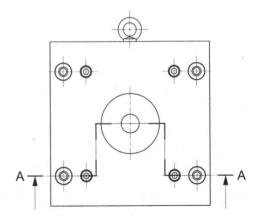

have specific file names and endings. Selecting the elements that you do not want
to section is a little more difficult in this case. Manufacturers present their models
on different levels of the international standard of quality 3D model description. It
is true that it is possible to describe a model in the STEP standard; however, there
is no specified minimum information block for a 3D model record that is to be used
for further activities in PDM systems. This is usually left to the individual users.
Large corporations usually set standards for the assembly and contents of data on
the element's or assembly's description.

Figure 15.28 shows this problem in connection with open menus and the procedure
for selecting a particular presentation. The clarity of presenting individual elements
is the key condition, and setting the structure of a product is of significant importance.
It is specified by the designer of a product. This is where the requirement appears for
the first time that prior to designing, the product structure should be created, based
on the functional structure of the product itself.

Presenting the elements and labelling the axes (centrelines) for each individual
element, and a proper hatching density, contribute to the explicitness and clarity of
the plan. Unfortunately, using default software settings results in significant illogical
and unclear issues on the cross-section part of the plan. The plan loses its message
and legibility and becomes a bunch of lines and characters. A simple example of a
plastic injection-moulding tool shows all the illogical and unclear issues that come
with it. It is shown in Fig. 15.29.

A proper representation and compliance with agreed standards (e.g., ISO) of a
multi-plane cross-section of the tool, as shown in Fig. 15.29, requires additional
labelling of the non-sectioned and sectioned parts, hatching density, and specifying the local axes that define the axiality of the elements. The most sensitive is the
hatching of sectioned surfaces of the same part, because in a good and clear representation, the hatching density for the same element is the same and of course identically

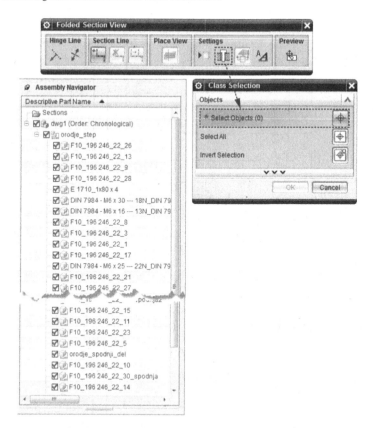

Fig. 15.28 Cross-section for a simple plastic injection-moulding tool and the problem of specifying the elements that are not to be shown as sectioned

oriented. All modelers usually have problems in these cases and occasionally perform representations rather unreliably.

The best example comes from comparing Figs. 15.29 and 15.30.

15.6 Parts List

The parts list is an important part of the documentation. It first appears in the assembly drawing. The parts list later appears as the fundamental information guide throughout the flow of information on a product. It represents the key information for PDM/PLM systems. The model gains its representation value, and appears in later phases of the PDM/PLM systems as a rough image of a product. The 2D manufacturing drawing remains at the level of the original information, defining the product in its details, and representing a high-quality industrial property.

A-A

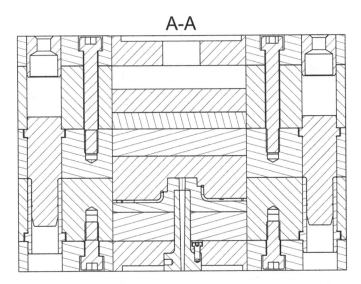

Fig. 15.29 Multi-plane cross-section of a plastic injection-moulding tool as a direct result of the modeller's default settings. The drawing is unclear and even misleading in some details. See Fig. 15.27 for details

A-A

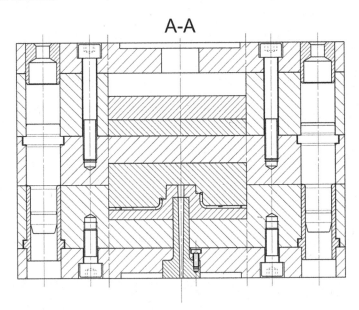

Fig. 15.30 Multi-plane cross-section of a plastic injection-moulding tool, following the rules of the ISO standard representation.See Fig. 15.27 for details

The parts list includes a collection of important data on a product. They are not created at once but they grow slowly. Some modelers provide an option at the beginning to enter the product data as "properties". It is wrong to think that all the product data are available at the beginning.

Data are collected in a logical sequence, together with a growing knowledge about the product:

1. Product name
2. Number of parts built in the assembly
3. Rough dimensions of the product in the assembly. Later, once the manufacturing drawing of individual elements has been defined, they can be changed or modified. They are usually written next to the product name, in the same column.
4. Position number (suitable for creating the building logic, i.e., the product assembly), directly related to the product ident number.
5. A rough definition of the product's material. It is changed or complemented through the analyses of stress, deformation, temperature and manufacturing technology.
6. Standard or ident number of the document, specifying the element on the basis of the manufacturing drawing or sub-assembly. Key information for relations in the PDM/PLM system.
7. Product weight, defined according to precise dimensions on the basis of the product's manufacturing drawing and can be defined only once the manufacturing drawing has been created
8. Annotation, defining the specifics of the element or assembly, presented by the drawing

The parts list normally does not include any descriptions of the specific details. Out of the eight groups of data, it should be noted that only numbers 1, 4 and 2 represent the initial data that can provide reliability and do not change throughout the assembly development. Only after the first approximation of precise dimensions can you write data under numbers 3, 5 and 6. The assembly weight is defined once all the elements have been specified and the standard parts clearly defined.

The standard part requires a special definition. Standard parts used to be defined by the standards of a state or a group of states. Nowadays, in the global world, there are the standards of global corporations, providing standard parts, referred to as purchasables. It is vital for a designer that the element is standard in order to provide adequate functional quality and its shape is so defined through product geometry that it can be included in the assembly. Describing the quality is prescribed by the technical report, which later appears as an additional description of the element with detailed characteristics.

For reasons of procedural clarity, let us take a look at completing a parts list that can later be translated into the PDM/PLM information system. NX software will be used for the presentation.

In the *Parts List* menu, NX allows the automated creation of the parts list (Fig. 15.31). The parts list has three categories in its basic setting: (1) position number, (2) part name (for the name, it takes the model's file name) and (3) number of

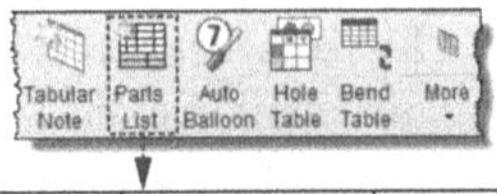

Fig. 15.31 Parts list layout
as pre-set by the software

12	SPACING_SLEEVE	1
11	HEX HEAD SCREW, GRADE C, DIN EN 24018,M10X30	1
10	HEX NUT, 1 A, FINE THD, DIN,M30X2	1
9	SAW_BLADE_HOL DING_RING	2
8	WASHER	1
7	PULLEY	1
6	PARALLEL_KEY	1
5	HOUSING	1
4	6210-2RS1_2_03_S TP	2
3	RETAINING RING ABOVE 25, DIN,90X3	1
2	RETAINING RING 10 TO 300, DIN,50X2	1
1	SHAFT	1
PC NO	PART NAME	QTY

pieces of each position. These three columns are the original information. You can see that these three columns do not follow the ISO standard order ((1) Position, (2) Number of pieces, (3) Name and dimensions). By moving the columns in the correct position, the software would significantly improve its usefulness. A parts list without pre-set (labelled) positions usually opens upon activation. The software can set the position numbers automatically.

The parts list should first be inserted onto the drawing plane, and only then are additional positions determined with the (*Auto Balloon*) command. The menu choice (Fig. 15.32) in the *Auto Balloon* sub-menu opens a new window that allows modifying positions. The first pre-set parts list has three columns. It should be noted that a new position is automatically adopted and can be moved from the parts list onto the drawing plane and the product, and vice-versa.

In this part, an NX modeller does not follow ISO standards, as shown in Fig. 15.33. The particular modeller does not label all the positions, although they are shown in the drawing, and the shape of the position lines is also not correct (ISO standard).

The relation between the parts list and the position numbers is fixed and built in each modeller. The pre-set layout of the parts list, which is inserted with the *Parts List* tool, can be manually modified. Besides a different number of lines, defined according to the number of positions, you should also add columns, according to data acquisition. It should be noted again that the parts list should be complemented in accordance with defining the elements on the plan, and not at the end, when the plan is completed and developed in all its details. It allows simultaneous product development control by means of the product structure, defined by the morphological matrix and by breaking the product down into partial functions. New columns represent additional vital information: (4) numbers of manufacturing drawings (or standard, if standard parts are used for the development), (5) material and (6) the

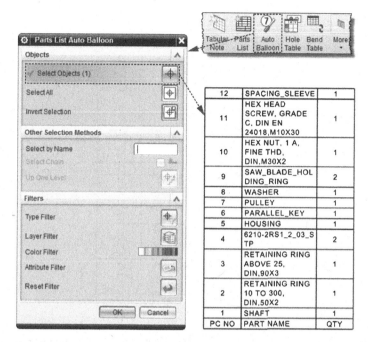

Fig. 15.32 Using the *Auto Balloon* menu

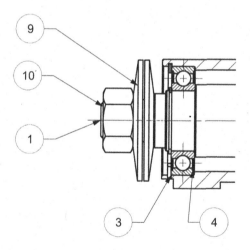

Fig. 15.33 Labelling positions in the *Auto Balloon* menu has some weaknesses for direct use

element's mass. Add a new column by marking an existing column, right-clicking and selecting *Insert > Columns to the Left/Columns to the Right* (Fig. 15.34).

Information can be added to the columns manually, or you can set or prescribe in advance what categories to enter in which column. These categories should first

Pos.	Title and dimensions	QTY
12	EEVE	1
11	DE C	
10		1
9	HOLDI	2
8	WASHER	1
7	PULLEY	1
6	PARALLEL_KEY	1
5	HOUSING	1
4	6210-2RS1_2_03_ST P	2
3	RETAINING RING ABOVE 25, DIN,90X3	1
2	RETAINING RING 10 TO 300, DIN,50X2	1
1	SHAFT	1

Context menu overlay:
- Select from List...
- Settings
- Insert → Columns to the Left / Columns to the Right
- Resize
- Select ▶
- Cut   Ctrl+X
- Copy   Ctrl+C
- Delete   Ctrl+D
- View ▶

Fig. 15.34 Adding new columns into the parts-list to allow extended data entries for each element (columns from 4 to 6 and beyond)

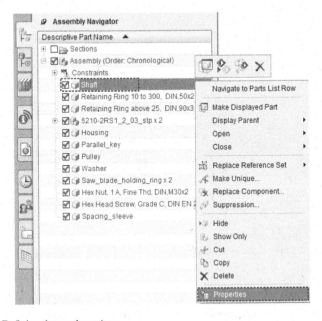

Fig. 15.35 Defining the part's settings

Fig. 15.36 The menu for adding new attributes

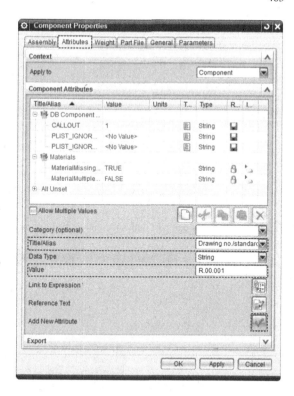

Fig. 15.36 The menu for adding new attributes

be defined at the level of the elements of a particular assembly. In the *Assembly Navigator* right-click on a piece and select *Properties*, Fig. 15.35.

By selecting *Properties*, a new window opens, where you can add individual attributes (Fig. 15.36). Write the attribute name (e.g., Drawing no./Standard) in the *Title* sub-menu, and its number (e.g., R.00.001) in the other, *Value* sub-menu. The number is presented as a value. Confirm your decision by clicking on the green ticked box and add your attribute to the parts list's data collection. By doing so, the drawing number is determined. Links to other positions and specifying automated procedures are included in the software. Example: specify in the parts list which column should record the "Drawing no./Standard" attribute. The number of the shaft's manufacturing drawing will be automatically recorded and added in this column. To specify the column where the drawings' numbers will be recorded, you first need to move the cursor to the location and then right-click on it and select the *Settings* command, Fig. 15.37.

After executing this operation, a settings window will open. Using the *Column>Attribute name* command, define "Drawing no./Standard", (Figs. 15.38 and 15.39). Using this command, all the drawings' numbers—previously defined in the procedure as explained above—will be copied into the selected column in the parts list.

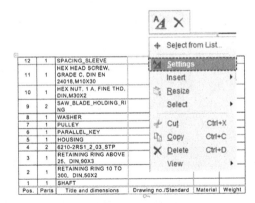

Fig. 15.37 Extended parts list with six columns, and setting attributes for each column

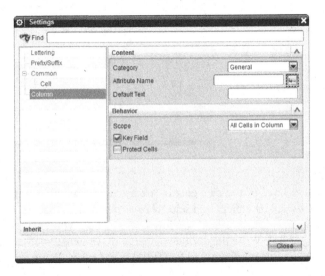

Fig. 15.38 The settings window as it appears in the *Column* sub-menu, and a marked click on the *Attribute Name* command

The element's title and dimensions should be specified in the model's file name. For clarity reasons—to have as clearly defined links as possible—in this case you also need to change the name of the column in the parts list (Fig. 15.40).

For the reasons of easier and more transparent work, it makes sense for the information on material and mass to be transferred into the parts list automatically. In this case, material should be prescribed at the element level, by a succession of *Tools > Material > Assign Materials* commands. Due to uniquely identified links, you need to define the material and mass attributes in the corresponding columns of the parts list, as shown on the example of entering the drawings' numbers. The parts list is

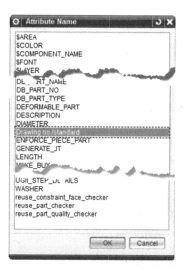

Fig. 15.39 Adding the "*Drawing no./Standard*" attribute into the column of choice

12	1	Spacing sleeve ⌀ 58/50x10	R.00.006	St 37.2	0.05
11	1	Screw M10x30	DIN 931	20898	0.03
10	1	Key 12x8x54	DIN 6885-1	10025	0.05
9	1	Retaining ring ⌀ 90x3	DIN 472	10027-1	0.04
8	2	Retaining ring ⌀ 50x2	DIN 471	10027-1	0.01
7	2	Bearing ⌀ 90/50x20	SKF 6210-2RS1		0.47
6	1	Nut M30x2	DIN EN 28674	8.8	0.24
5	1	Washer ⌀ 60x8	R.00.005	St 37.2	0.16
4	2	Saw blade holding ring ⌀ 100x8	R.00.004	St 37.2	0.40
3	1	Pulley ⌀ 125x63	R.00.003	AlMgSi	1.33
2	1	Housing ⌀ 101,6x300	R.00.002	St 37.2	7.52
1	1	Shaft ⌀ 60x450	R.00.001	St 37.2	7.92
Pos.	Parts	Title and dimensions	Drawing no./Standard	Material	Weight

Projections: ⊟ ⊕	General tolerance principles: ISO 8015	Scale: 1:2.5		Weight: 17.3 kg
			Date	Name	Title and dimensions:
		Drawn	dd.mm.y	Zorko D.	
		Chk'd	dd.mm.y	Demsar	**Shaft for saw**
		Std. C.	dd.mm.y	Duhovnik	⌀ **125x140x464**
		University of Ljubljana Faculty of Mechanical Engineering	Drawing number: R.00.000	Sheet 1 / 1 S.
Sign	Change	Day	Name Dat.: risba angleska sestavna popravljena

Fig. 15.40 A completed parts list, having the same length as the head of the drawing or the plan is positioned above the title block on the assembly drawing

12	1	Spacing sleeve ⌀ 58/5	⊕ Select from List...		006	St 37.2	0.05
11	1	Screw M10x30			931	20898	0.03
10	1	Key 12x8x54	⌖ Hide Ctrl+B		385-1	10025	0.05
9	1	Retaining ring ⌀ 90x3			472	10027-1	0.04
8	2	Retaining ring ⌀ 50x2	▦ Edit...		471	10027-1	0.01
7	2	Bearing ⌀ 90/50x20			0-2RS1		0.47
6	1	Nut M30x2	▥ Settings		28674	8.8	0.24
5	1	Washer ⌀ 60x8	▥ Cell Settings...		005	St 37.2	0.16
4	2	Saw blade holding ring	⬚ Associate to View		004	St 37.2	0.40
3	1	Pulley ⌀ 125x63			003	AlMgSi	1.33
2	1	Housing ⌀ 101,6x300	▢ Disassociate from View		002	St 37.2	7.52
1	1	Shaft ⌀ 60x450	Select ▸		001	St 37.2	7.92
Pos.	Parts	Title and dime			Standard	Material	Weight
Projections: ◁▭ ⊕		General to	⇅ Sort...		2.5		Weight: 17.3 kg
			⬚ Export...		ensions:		
			▨ Edit Using Spreadsheet		Shaft for saw		
			▨ Edit Without Spreadsheet		⌀ 125x140x464		
			⬚ Update Tabular Note		ber: R.00.000		Sheet 1
			✂ Cut Ctrl+X				1 S.
Sign	Change	Day	Name D	⬚ Copy Ctrl+C			

Fig. 15.41 Exporting data from a completed parts list to either a .txt or .xls file

	A	B	C	D	E	F
	▦ Delovni list v Tabular Note					
1	12	1	Spacing sle	R.00.006	St 37.2	0.05
2	11	1	Screw M10	DIN 931	20898	0.03
3	10	1	Key 12x8x5	DIN 6885-1	10025	0.05
4	9	1	Retaining r	DIN 472	10027-1	0.04
5	8	2	Retaining r	DIN 471	10027-1	0.01
6	7	2	Bearing <O	KF 6210-2RS		0.47
7	6	1	Nut M30x2	IN EN 2867	8.8	0.24
8	5	1	Washer <O	R.00.005	St 37.2	0.16
9	4	2	Saw blade	R.00.004	St 37.2	0.40
10	3	1	Pulley <O>	R.00.003	AlMgSi	1.33
11	2	1	Housing <C	R.00.002	St 37.2	7.52
12	1	1	Shaft <O>6	R.00.001	St 37.2	7.92
13	Pos.	Parts	and dimensing no./Star		Material	Weight

Fig. 15.42 Parts list, imported to a special file in the .xls format

completed, as shown in Fig. 15.40, once the data on mass and material have been entered.

In order to be able to re-use the layout of the parts list, you can save it as a template and use it again for other assemblies. The importance of the parts list should be understood in particular in connection with the PDM/PLM system of information. It results in integration and the reliable transfer of all the data. Another suitable format for data transfer to other information systems is the .txt file format. Data transfer to this format is executed with a right-click and the cursor on the marked

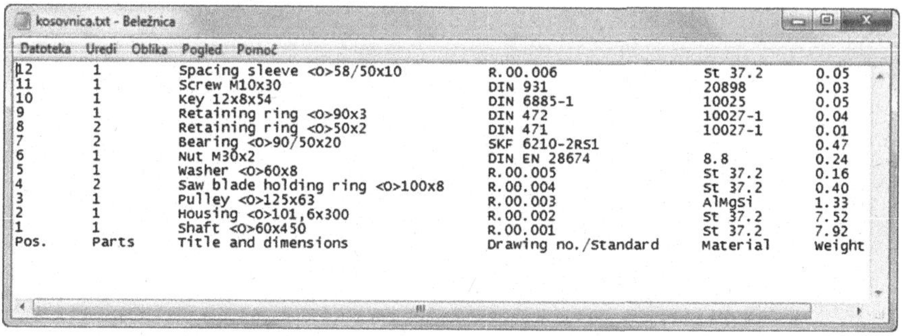

Fig. 15.43 Parts list, imported to a special file in the .txt format

parts list and using the **Export** command. The parts list can also be exported to the Excel file by right-clicking on the marked parts list and selecting the **Edit Using Spreadsheet**command. Both transfers are shown in Fig. 15.41.

For explicitness and most of all for the purpose of useful forms for adding additional information, the .xls file format is very suitable for further processing. A file should be re-designed in the sense that the columns are legible and show all the characters in the visible field. Figure 15.42 shows a simple data transfer as an example of fixed column width. Figure 15.43 shows export to a .txt file.

References

1. Agoston, M. K. (2005). *Computer graphics and geometric modelling: Mathematics*. Berlin: Springer.
2. Anand, V. D. (1993). *Computer graphics and geometric modeling for engineers*. Hoboken: Wiley.
3. Bertoline, G. R., Wiebe, E. N., Miller, C. L., & Nasman, L. O. (1995). *Technical graphics communication*. IRWIN.
4. Boyer, E. T., Meyers, F. D., Croft jr., F. M., Miller, M. J., & Demel, J. T. (1991). *Technical graphics*. New York: Wiley.
5. Brunet, P., Hoffman, C., & Roller, D. (2000). *CAD Tools and Algorithms for Product Design*. Berlin: Springer.
6. Divjak, S. (2000). *Računalniška grafika*. Fakulteta za računalništvo in informatiko.
7. Duhovnik, J. (2003). Techniques and methods for product developments. In *IV Wroclawskie Sympozjum: Automatyzacja produkcji* (pp. 93–101). Wroclaw, 11–12 Dec 2003.
8. Duhovnik, J., & Demšar, I. (2010). *Space modelling: exercise book—course book*. Ljubljana: Fakulteta za strojništvo.
9. Duhovnik, J., & Demšar, I. (2012). *Modeliranje Prostora I: zbirka vaj*. Ljubljana: Delovniučbenik za NX. Fakulteta za strojništvo.
10. Duhovnik, J., Kljajin, M., & Opalić, M. (2009). *Inženirska grafika*. Ljubljana: Fakulteta za strojništvo.
11. Duhovnik, J., Kušar, J., Tomaževič, R., & Starbek, M. (2006). Development process with regard to customer requirements. *Concurrent Engineering*.
12. Duhovnik, J., Žargi, U., Kušar, J., & Starbek, M. (2009). Project-driven concurrent product development. *Concurrent Engineering*.
13. Earle, H. J. (2002). *Graphics for engineers* (6th ed.). Texas: Texas A&M University.
14. Farin, E. G. (2002). *Curves and surfaces for CAGD* (5th ed.). Waltham: Academic Press.
15. Foley, J. D. (1993). *Computer graphics: Principles and practice*. Boston: Addison Wesley.
16. Glodež, S. (2005). *Tehnično risanje*. Ljubljana: Tehniška založba Slovenije.
17. Hoischen, H., & Hesser, W. (2005). *Technisches Zeichnen*. Berlin: Cornelsen.
18. Kljajin, M. (1991). *Tehničko crtanje*. Slavonski Brod: Strojarski fakultet u Slavonskom Brod.
19. Koludrović, Ć. (n.d.). *Tehničko crtanje u slici s osnovnim vježbama*. Beograd: Naućna knjiga.
20. Kovać, B. (1967). *Tehničko crtanje—Priručnik za kovinsku struku*. Zagreb: Školska knjiga.
21. Kraut, B. (2002). *Krautov strojniški priročnik*. Ljubljana: Littera Picta.
22. Matek, W., Muhs, D., Wittel, H., Becker, M., & Jannasch, D. (2000). *Roloff / Matek Maschinenelemente*. Wieseaden: Vieweg Verlag.

J. Duhovnik et al., *Space Modeling with SolidWorks and NX*,   489
DOI: 10.1007/978-3-319-03862-9, © Springer International Publishing Switzerland 2015

23. Mortenson, M. E. (2007). *Geometric transformations for 3D modeling.* South Norwalk: Industrial Press.
24. Opalič, M., Kljajin, M., & Sebastjanović, S. (2007). *Tehničko crtanje* (2nd ed.). Sveučilište u Zagrebu: Sveučilište u Slavonskom Brodu.
25. Pomska, G. (1986). *3D-Grafik auf dem PC: Modellierung, Projektion.* Vogel-Buchverlag: Sichtbarkeit.
26. Prebil, I., & Zupan, S. (2011). *Tehnična dokumentacija.* Ljubljana: Tehniška založba Slovenije. (multiple editions).
27. Ren, Z., & Glodež, S. (2003). *Strojni elementi—I.del.* Maribor: Založništvo fakultete za strojništvo.
28. Siemens PLM Software. (2010). *NX Sheet Metal Design—NX 7.5—Student Guide.*
29. Siemens PLM Software. (2010). *Routing Mechanical—NX 7.5—Student Guide.*
30. Siemens PLM Software. (2010). *Sketcher Fundamentals—NX 7.5—Student Guide.*
31. Stephen, M., & Samuel, P. E. (2013). *Basic to Advanced Computer Aided Design Using NX8.5.* Design Visionaries.
32. Stroud, I. (2006). *Boundary representation modelling techniques.* Berlin: Springer.
33. Valentino, J., & DiZinno, N. (2010). *SolidWorks for technology and engineering.* New York: Industrial Press.
34. Verlinden, J., & Horvath, I. (2008). *Enabling interactive augmented prototyping by a portable hardware and a plug-in-based software architecture* (pp. 458–470). Ljubljana: Journal of Mechanical Engineering.
35. Viebahn, U. (2004). *Technisches Freihandzeichnen* (5th ed.). Berlin: Lehr- und Übungsbuch. Spreinger.
36. Watt, A. (1990). *Fundamentals of three-dimensionals computer graphis.* Boston: Addison Wesley.
37. Zupan, B. et al. (2007). *SolidWorks 1, uporabniški priročnik, User manual 1.* ib-CADdy.
38. Zupan, B. et al. (2007). *SolidWorks 2, uporabniški priročnik, User manual 1.* ib-CADdy.

Printed in the United States
By Bookmasters